C. CLAIBORNE RAY

The New York Times Second Book of Science Questions and Answers

C. Claiborne Ray has been an editor for *The New York Times* for twenty-five years, and has been the writer of *The New York Times* Science Q&A column since 1988. A resident of a historic Brooklyn neighborhood, she is a cat lover and a major-league jazz fan.

Victoria Roberts, a distinguished cartoonist, illustrates *The New York Times* Science Q&A column weekly. Her work regularly appears in *The New Yorker*.

Henry Fountain is an editor at *The New York Times*.

The New York Times

SECOND BOOK OF

Science
Questions
and
Answers

The New York Times

SECOND BOOK OF

Science

Questions *and*

Answers

*225 New, Intriguing, and Just Plain
Bizarre Inquiries into Everyday
Scientific Mysteries*

C. Claiborne Ray
Drawings by Victoria Roberts
Edited by Henry Fountain

ANCHOR BOOKS
A Division of Random House, Inc.
New York

FIRST ANCHOR BOOKS EDITION, APRIL 2003

Text copyright © 2002 by The New York Times Company
Illustrations copyright © 2002 by Victoria Roberts

All of the questions and answers have been previously published in
The New York Times.

"The Turtle" by Ogden Nash is reprinted by permission of the Estate of Ogden Nash.

Library of Congress Cataloging-in-Publication Data
Ray, C. Claiborne.
The New York Times second book of science questions and answers :
225 new, intriguing, and just plain bizarre inquiries into everyday scientific
mysteries / C. Claiborne Ray ; edited by Henry Fountain ; drawings
by Victoria Roberts.—1st Anchor Books ed.
p. cm.
Includes bibliographical references.
ISBN 0-385-72258-3 (pbk.)
1. Science—Miscellanea.   I. Fountain, Henry.   II. New York Times.
III. Title.
Q173 .R394 2003
500—dc21   2002026192

Book design by Debbie Glasserman

www.anchorbooks.com

Printed in the United States of America
10 9 8 7 6 5 4 3 2 1

Contents

Preface

Back when I was in college, a favorite bit of graffiti scrawled on bathroom stalls all over campus posed the age-old question: What is reality?

That must be one of the few questions that has never been asked of C. Claiborne Ray, who since 1988 has been responding to queries about science from *New York Times* readers every week in the paper's Science Times section.

Although the question has never been asked, Claiborne answers it all the time. For making sense of science is, at its heart, making sense of the real world. So when readers write to ask if the universe is rotating, or what makes the holes in Swiss cheese, or how high a housefly can fly, Claiborne's answers build up their—and our—understanding of reality, bit by bit.

Times readers are curious. They want answers to easy questions, like why plants grow straight up, and to questions that have stumped the experts for ages, like whether a guinea pig is a rodent. They want to know about creatures large (whales), small (dust mites), and in between (bears and chipmunks).

They want to know all about themselves, too: why their feet hurt and their eyes are sore, and what that thing dangling in the back of their throats is. They want to know if they can live exclusively on pizza—though why they want to know this is beyond me—and why they can't keep their eyes open after lunch.

They are curious about worlds beyond Earth. They want to know, for instance, if Pluto will ever run into Neptune, or if it's possible that alien civilizations are watching *I Love Lucy* reruns as we speak. And occasionally they ask a famous epistemological question, like: How many angels can dance on the head of a pin?

But mostly they hunger for knowledge about our world, about AIDS and the common cold, about clouds and trees and rivers and ice and hundreds of other things—225 things in total in this book.

Claiborne takes on all comers. In this she is aided by experts who are kind enough to share *their* understanding of reality. Those who aided Claiborne are acknowledged in the Notes on Sources at the end of this book.

There are no questions left unanswered in this, the second volume of *The New York Times Book of Science Questions and Answers*. But there are plenty of questions left unasked. Readers seeking answers to those will have to wait for future volumes. Or perhaps the responses that Claiborne provides here will inspire readers to go out and find some answers of their own.

— HENRY FOUNTAIN

The New York Times

SECOND BOOK OF

Science
Questions
and
Answers

CHAPTER 1

Reaching for the Stars

BIRTH AND DEATH

Q. Are stars all burning out, or are new ones forming?
A. Stars are being born as well as dying, but the rate varies greatly from galaxy to galaxy.

Stars form from huge clouds of dust and gas. If a cloud begins to contract because of its own gravity, its interior heats up as gravitational energy is converted to heat energy, reaching millions of degrees, and nuclear reactions begin that change one element into another, releasing energy.

The pressure tends to expand the cloud back out, but eventually equilibrium is reached. That is essentially what a star is—a mass of gas at equilibrium between inward pressure from gravity and outward pressure from nuclear reactions.

A star has a finite lifetime because it is burning fuel. For 90 percent of its life, it burns hydrogen into helium. When the hydrogen is used up, the pressure decreases, but gravity never disappears, so the star contracts until the temperature climbs again, this time reaching hundreds of millions of degrees, while reactions convert helium to carbon and oxygen. The star can then remain stable for a briefer time. Eventually the star dies, when the reactions no longer produce energy, but only consume it.

IN A SPIN

Q. Is the universe rotating?
A. Most astronomers would say no. There is no known mechanism that would give the universe so much angular momentum, or spin, at its beginning, and few mechanisms for adding spin later.

To know for sure if the universe rotates, scientists would need to know the velocities of millions of galaxies, over all regions of the sky and out to very great distances. Analyses of these velocities would be necessary to see if they indicated a common center about which galaxies were rotating, and a sense of direction on average.

Since it may be a while before the velocities of millions of galaxies are known, astronomers are trying to answer a simpler question: Are there regions of space toward which large numbers of galaxies are moving as a group?

Q. What would kill you if you fell into a black hole?

A. You might not die right away, but you would eventually be pulled apart by the force of gravity. As you fell in and even afterward, you might not lose consciousness, but the pull of gravity on your feet would be stronger than on your head, and you would be stretched, then torn apart. The difference in force is called the tidal force and is like that in the ocean, except in more extreme form. The force would be less if it was a big enough black hole; in a small black hole it might kill you before you disappeared beneath the event horizon, the edge of the hole.

But even in a larger black hole, the tidal force always gets you in the end. Once you fall in, you can't avoid falling toward the center, and the force would kill you before you reached the center. How long it would take depends on how big the black hole is. If it was big enough so that the tidal force didn't kill you before you fell in, you might have an hour or several hours before being torn apart. In a small one, such as one that forms when a star collapses, you wouldn't have much time, perhaps a thousandth of a second. That would happen with a run-of-the-mill black hole, like those that might be found in the Milky Way Galaxy.

Averting Your Eyes

Q. When I look at a constellation, I can see the fainter stars better out of the corner of my eye. Why?

A. Because the eye has two kinds of receptors, cones for fine resolution and color and rods for dim light, and the rods tend to be located around the periphery, for viewing the edges of the field of vision.

Cones are extremely good at high definition and for precise positioning of pinpoints of light. Rods don't give nearly as fine resolution and don't distinguish colors, but are much more sensitive. In effect, rods serve as night-vision sensors.

The result is that when you go out at night, you can clearly see the bright stars and planets at the center of your eye, but you can see the fainter ones out of the corner of your eye. Astronomers call this averted vision. For example, if they look with the unaided eye straight at a galaxy they know is there, they may not see it, but if they look off to one side, they can easily see the fuzzy gray patch that is the galaxy.

Cameras, too, tend to give up resolution and sensitivity to color when they achieve better sensitivity to low light levels.

SEEING STARS

Q. Why can't you see the stars in photos or videos taken by astronauts?

A. Such pictures do not ordinarily show stars because the stars are not bright enough in comparison to the nearby Sun and the things it shines on.

Virtually all of the astronaut photos are of objects brightly illuminated by the Sun. To capture them on film without overexposing the image, you need a relatively short exposure, which does not provide enough time for the film to capture images of stars. If there is no other strong light source in the picture, however, a photo can show stars.

A SOUTH STAR?

Q. What is used as a pole star in the Southern Hemisphere, where navigators can't see the North Star?
A. The closest thing to a south star for navigators south of the equator is a pair of stars in the Southern Cross, Crux Australis (or just Crux to astronomers). Alpha Crucis (its brightest star) and Gamma Crucis (the third brightest) point almost straight to the south celestial pole.

The striking Southern Cross, which has four stars brighter than the second magnitude in a cross or kite shape, takes up only 68 square degrees of the sky. The area defined by the Southern Cross has other interesting features, notably the Coalsack Nebula, which looks like a round blank spot in the sky but is actually a cloud of gas-laden dust that blocks the background starlight, and the Jewel Box cluster, designated NGC 4755, an open cluster of more than 100 stars surrounding Kappa Crucis, which is bright red in a modest telescope.

HELLO OUT THERE!

Q. In the search for extraterrestrial intelligence, or SETI, what kinds of signals would be considered evidence of transmission by intelligent life? How would scientists respond?

A. Such a transmission might take many forms, but would probably encode mathematical formulas. The reply would depend on the content; it would not be made by scientists, but would come after extensive international consultation.

As for verification, the main feature distinguishing signals produced by a transmitter from those produced by natural processes is their spectral width—that is, how much room on the radio dial they take up. As far as scientists know, any signal less than about 300 hertz wide is artificially produced.

Other telltale characteristics might be coded information, like the message beamed from the Arecibo telescope in 1974, which included data like the senders' location (third rock from the Sun, in the case of Earth). Another important test would be a confirming observation of the same signal at another radio telescope.

Once confirmed, the discovery would be announced based on a plan set up by six international space agencies. First, the scientific community would be notified through the International Astronomical Union and the United Nations. Then international authorities would draft a reply.

SITCOMS IN SPACE

Q. How far out in space could Earth's television and radio signals be detected?

A. The speed of light can seem fairly slow when you're talking about communicating across galaxies. FM broadcasts and the earliest television programs from, say, half a century of broadcasting have reached a distance of 50 light-years, or about 294 trillion miles, from Earth.

The nearest star is about 4 light-years away, and there are on the order of several thousand stars within the 50-light-year range. So the earliest episodes of *I Love Lucy* are washing over a new star system at the rate of about one system a day.

Any civilization on the receiving end would need a very large antenna to pick up the broadcasts, about the size of Manhattan, but scientists suggest it could be done. But the chances that one star in a few thousand had some sort of civilization might be a big overestimate.

SHINE ON, HARVEST MOON

Q. Is the harvest moon brighter than other full moons?
A. Ordinarily, the mid-September harvest moon is no brighter than any other full moon, but it does provide more hours of moonlight.

In autumn, the orbital path of the Moon and its visibility in the sky combine so that it stays above the horizon for an unusually long time at the full moon and a day before and after.

The brightness of the Moon is determined by how much sunlight falls on it, which is pretty much constant; how much appears to be lit up from where we see it, from crescent to full; and how high it is in the sky. The higher the Moon is, the more light we see, because it has a shorter path through the dust and smog of the atmosphere.

In 1997, the harvest moon of September 15 was the brightest of the year, by a small margin, because it came within a few hours of perigee, the Moon's closest approach to Earth. The difference in brightness between the Moon's closest approach and farthest point amounts to 12 to 13 percent. It was also the largest full moon in angular size; its apparent diameter, normally about 30 arc-minutes, was 32 or 33 arc-minutes.

The varying Earth-Moon distance is why solar eclipses are sometimes total, when the Moon completely blocks the Sun, and sometimes annular, with a ring of light visible around the Moon.

ON THE THIN SIDE

Q. Did the Moon ever have an atmosphere?
A. It has one now, though it is a very thin and highly dispersed collection of molecules, not suitable for breathing by Earthlings.

The existence of a lunar atmosphere was reported in 1933, based on observation of the Moon using a mask that filtered out moonlight in order to study the spectrum of light emitted by sodium. Although sodium is believed to be just a trace in the Moon's atmosphere, it is studied because it is relatively easy to detect and is used as a marker for other components, such as potassium, neon, argon, and helium.

A 1993 study of lunar sodium by Boston University

scientists using improved instruments determined that the atmosphere extended at least 5,000 miles above the Moon's surface. The molecules, however, are few and far between, only an estimated 10 million per cubic centimeter near the Moon's surface; Earth's atmosphere is about a billion times as dense.

The sources of the atmosphere are believed to be the release of gases from within the Moon by moonquakes (a phenomenon called outgassing) and the loosening of molecules from the surface by the impact of molecules from the solar wind or by meteorites.

A few moons of other planets have much more impressive atmospheres, like that of Jupiter's Titan, a thick haze of nitrogen and methane, and that of Europa, a thin wisp of oxygen.

MARS COORDINATES

Q. How is zero degrees longitude, the equivalent of the Greenwich meridian, determined for Mars?
A. A small, well-defined crater named Airy-0, near the planet's equator, was designated as the starting point for the 360 degrees of Martian longitude.

The satellite Mariner 9 began photographing Mars on November 13, 1971, sending back thousands of detailed pictures of the planet's surface on which to base a map. Using the information captured by this mapping project, the scientists at the U.S. Geological Survey's Center of Astrogeology at Flagstaff, Arizona, were able to refine their calculations so that the line for the prime meridian goes through the exact center of the crater, which is about three-tenths of a mile in diameter.

The zero-degree crater was named Airy-0 in honor of George Biddell Airy, a British astronomer who lived from 1801 to 1892. He became Astronomer Royal in 1835, and among his many accomplishments was the building of the Transit Circle telescope in the Greenwich Observatory's Meridian Building in 1850. The cross-hairs in the eyepiece of the telescope define zero degrees longitude for Earth.

ROCKS FROM THE RED PLANET

Q. How do scientists determine that a rock on Earth came from Mars rather than someplace else?

A. The best idea scientists have of the geochemistry of Mars comes from the two Viking robots that landed on Mars in 1976. Findings from the robots' weeks of readings of things like the Martian atmosphere are compared with the chemical signatures of meteorites found on Earth.

The first such object confidently identified as Martian was reported in 1983. The object, a rock eight inches in diameter that weighed 17.5 pounds, had been picked up in 1979 at the Elephant Moraine near McMurdo Sound in Antarctica.

Trapped in bits of glass in the object, designated EETA 79001, were some so-called noble gases—neon, argon, krypton, and xenon—strikingly similar in abundance to those of the Martian atmosphere, as determined by the Viking missions.

There are now about two dozen meteorites thought to be from Mars.

PALE IN COMPARISON

Q. Why is Venus so much brighter in the sky than Mercury, which is closer to the Sun?
A. Venus is brighter because it is much larger, because it has a reflective atmosphere, and because it makes close approaches to Earth.

First, Mercury is about 3,100 miles in diameter, compared with about 7,700 miles for Venus.

Second, there is no atmosphere surrounding Mercury, while Venus has a thick atmosphere, composed chiefly of clouds of carbon dioxide. These clouds have a very high reflectivity, causing Venus to appear to shine with great brilliance.

Third, Venus can approach the Earth much more closely than Mercury. In fact, Venus periodically comes closer to Earth than any other known celestial body, except for the Moon, as close as 25 million miles. All of these things allow Venus to shine more brightly than not just Mercury, but than any of the other planets in our solar system.

Because Mercury and Venus are closer to the Sun than Earth is, they can show phases like those of the Moon. When Mercury shines at its brightest, it appears through a telescope as a gibbous or nearly full phase. When Venus appears to shine at its most brilliant, it appears in telescopes not as a nearly full phase, but as a crescent.

ON THE TRACK OF VENUS

Q. I have been told that the Maya made precise measurements of Venus. What did they learn?

A. It would be more accurate to say that the ancient Maya of Central America, who had no telescopes, did not measure Venus itself but kept careful records of its apparent track in the sky.

Because of the relative orbits of Earth and Venus, Venus is first invisible, then appears in the morning sky, then disappears, then reappears in the evening sky, then fades from view again.

An important part of the complex Mayan calendrical system is based on the 584 days it takes for Venus to make that complete cycle. The Venus cycle was meshed with a sacred calendar of 260 days and the 365-day solar calendar to come up with a cycle of about 2,920 days, or 8 solar years, in which the calendars came out approximately even.

COSMIC COLLISION

Q. Pluto sometimes crosses the orbit of Neptune, suggesting to me that they might collide at some point. Should I worry?

A. No. Their closest approach is not close at all—about the distance between Uranus and the Sun.

One reason they cannot pass any closer is that Neptune completes its orbit three times in the time that Pluto goes around twice. This exact relationship is called a resonance.

Another such relationship, called the Kozae resonance, provides an added protection. The orbit of Pluto is not in the same plane as Neptune's but is inclined to it at an angle of about 17 degrees. When Pluto is closest to the Sun, it is about as far away from Neptune's orbital plane as it can be, and there is close to the maximum separation when the two planets are crossing at the same distance from the Sun.

If the planets didn't have these protections, a collision would have happened long ago and we would never have known about Pluto's existence. Perhaps over hundreds of millions of years they could get out of resonance, but that is a long enough period that no one would have to worry.

Pluto is not alone in the Kozae resonance; at least fifty other objects orbiting the Sun, called Plutinos, enjoy the same protection.

THE PLANETARY PLANE

Q. Why are the orbits of the Sun's planets arranged in a more or less flat plane?

A. The planets are believed by many authorities to have coalesced from a cloud of gas early in the evolution of our solar system, beginning roughly 5 billion years ago. Since the gas cloud was disk shaped, when the planets came to be within this cloud, they all formed basically along the same plane.

There is a notable exception, the tiny planet Pluto, which has an orbit that is tipped as much as 17 degrees to the plane of the solar system. It has been hypothesized that Pluto might be everything from an escaped moon of Neptune to a special type of minor planet or asteroid.

Because of its anomalous orbit and other factors, there is great debate among astronomers as to Pluto's exact status as a member of the solar system.

RING AROUND THE SUN

Q. How does a coronagraph work? How could I build one?

A. A coronagraph is an optical device that blocks the light from the Sun's disk, making it possible to observe the corona, the very thin and hot upper level of the solar atmosphere, at the edge of the disk. A lens focuses an image of the Sun onto a masking or occulting disk that prevents the light from proceeding farther into the telescope.

A coronagraph requires very high quality optics assembled in a dust-free atmosphere. Near sea level, a coronagraph would be virtually useless, because the scattered light from the Earth's atmosphere would overwhelm light from the corona. Scientists put coronagraphs high in the mountains or launch them into space.

But any telescope can be turned into a coronagraph, simply by taking it to observe a total solar eclipse. (Just remember, for safety, never to look at the eclipse directly; project the image out through the eyepiece onto a piece of paper.) Because of the happy coincidence of the Moon's size and distance, when the Moon passes directly between the Sun and the Earth, it is usually in just the right position to make a very effective occulting disk. And during an eclipse, the Earth's atmosphere is also in the Moon's dark shadow, so light scattering is not a problem.

In fact, scientists still find a total solar eclipse to be an unexcelled opportunity to study the Sun's corona.

SPACE TRAFFIC CONTROL

Q. Is there an international agency that coordinates satellite launchings?

A. The United Nations keeps a registry and a data center at NASA assigns designations, but only communications satellites are subject to practical controls.

Every satellite must be registered to a particular country, for reasons of liability under space law. A registry has been maintained by the UN Secretariat since 1962, and another registry, established under a 1976 agreement, is kept by the agency's Committee on the Peaceful Uses of Outer Space.

Registry listings include a designation, determined by the World Data Center for Rockets and Satellites at NASA/Goddard Space Center, for each satellite. The listings include details of each object's orbit, altitude, and function.

But only the International Telecommunication Union coordinates launchings, by assigning the frequency slots for transmissions so that satellites won't interfere with one another.

STEERING THE SHUTTLE

Q. Why does the space shuttle roll after launching? How does it catch up to the space station without bumping into it?

A. Shortly after the shuttle clears the launching tower, the software of its automated guidance, navigation, and control system initiates a roll command to line it up on the desired course and heading. The roll program tells the engine nozzles how much to gimbal, or swivel, to aim the shuttle properly. It goes straight up for only a short time, then rolls to take advantage of Earth's rotation under it to get into orbit.

The shuttle goes downrange to get into orbit, actually heading downrange much faster than it goes up in altitude.

As for catching up to the space station, it is useful to consider the analogy of runners on a track. The station is in the outside lane, its orbit 240 miles above Earth, where it takes 92 minutes to go around the world at 17,500 miles an hour. The shuttle, in the inside lane at a lower altitude, takes only 90 minutes a lap. Over two days, its altitude is slowly raised until it matches that of the station. With both vessels traveling at the same orbital speed, the impact speed is about an inch per second.

SCI-FIDELITY

Q. Why is it that video from the space shuttle is usually as clear as most TV pictures, while audio is nearly as scratchy as transmissions from the sixties?

A. It has largely to do with the fidelity of the microphones, not the quality of space-to-ground communications systems. The astronauts wear wireless mikes that are as light and unobtrusive as NASA can make them. An FM signal goes from the astronaut to a communications device, then is radioed through a satellite system to the ground.

The microphones are fairly low fidelity. The better they are, the heavier they are, and on the shuttle, every ounce counts. NASA is not trying to obtain broadcast quality, and as long as the astronauts can be understood, that is enough.

Astronauts are given instruction in techniques for using a microphone, like not talking directly into it and avoiding the explosive *t*s and *b*s that are likely to make the mike pop. But they have got other things to worry about, and proper mike procedure is far down on the list.

GUNSMOKE IN ORBIT

Q. If I write a murder mystery set in space, can I have one of my characters kill another by firing a gun?

A. Yes, a gun fired in space or on a space station or rocket ship should work as well as it would on Earth or possibly even better.

The laws of physics still apply in space. The path of the bullet should be fairly straight, as gravity is greatly reduced, though not entirely absent.

There may be a reduced air supply on a spaceship, and there would certainly be one outside it, so the bullet would encounter less friction and would probably continue on its path longer.

Newton's third law, that for every action there is an equal and opposite reaction, means there would still be a recoil from the explosive expellation of the bullet.

As for the explosion that gets the bullet going, an old-fashioned six-shooter should still work very well, even in the relative absence of air. Gunpowder contains its own oxidizing agent, saltpeter, formally known as potassium nitrate, or KNO_3.

Some problems present themselves for the shooter. If he is aboard a spaceship, a bullet hole could cause the loss of his own air supply; he should probably wait for a space walk.

LOST IN SPACE

Q. Astronauts who stay in space for a long time lose muscle and bone mass. Is it permanent?

A. There is growing evidence that exercise can help astronauts recover from muscle and bone losses and limit such losses.

In the near absence of gravity, bone mass decreases in the load-bearing regions of the skeleton, and muscles shrink when not used. But Michael Foale, who was on

the Russian space station Mir for four and a half months, said that he had been able to recover most of the bone density and all of the muscle he lost on Mir through weight lifting, running, and swimming. Muscle recovery took about six months; bone rebuilding, to up to 99 percent of his preflight levels, took a year.

After earlier long flights, astronauts often had trouble recovering. On the 184-day Skylab 4 mission in 1974, astronauts lost 4.5 percent to 7 percent of their bone mass and were still down about that much 5 to 7 years later.

When both Shannon Lucid and Norman Thagard spent long periods aboard Mir, he lost up to 11.7 percent of his bone mineral and 17.5 pounds of overall muscle and weight, while she maintained her weight and lost less muscle mass and bone mineral than he did, even though she was in space 73 days longer. Dr. Lucid exercised more; she put in hundreds of hours on Mir's treadmill and stationary bicycle.

GETTING HOME FROM MARS

Q. Once people finally get to Mars, how will they ever get back?

A. The current plans involve bringing along a rocket that could be launched from Mars into a low orbit, where it would meet another craft that would return to Earth. The vehicle that returns to Earth would carry some kind of Earth-entry vehicle, like a capsule or shuttle plane.

To make the return cheaper by bringing less material from Earth, the Martian atmosphere would be used to produce much or most of the fuel to get the rocket from Mars into Mars orbit.

There are two alternative plans for providing fuel. In the first version, the travelers would carry along some fuel—methane, propane, or hydrazine—to burn with oxygen from the carbon dioxide that makes up most of the Martian atmosphere.

In the other scheme, both the carbon and the oxygen from the carbon dioxide would be used. The carbon would be combined with hydrogen carried from Earth to make methane to be burned with the oxygen.

An unmanned spacecraft might be sent to Mars to make the fuel even before humans leave Earth.

CHAPTER 2

Down to Earth

FINDING THE CENTER

Q. What is meant by the term *geographic center*? How is it determined? What is the geographic center of the United States? Is that with or without Hawaii and Alaska?

A. What the United States Census Bureau calls the geographic center of area is the point on which the surface of a geographical entity would balance if the surface were a plane of uniform weight per unit of area.

For the United States, that point has been in Butte County, South Dakota, on the state's western border, since the census of 1960, the year the bureau began including Hawaii and Alaska in its calculations. If only the forty-eight contiguous states are included, the point is in Smith County, Kansas, which is roughly in the middle of the state's northern border.

The Census Bureau also determines the population center of the United States at each census. The statistical method it uses to do this assumes that the map of the United States is flat, weightless, and rigid; the population center is the point on which this surface would balance if weights of equal value were placed on it, with each weight representing the location of one person on a particular date.

For the 2000 census the point lies in Crawford County, Missouri, about 90 miles southwest of St. Louis.

FROM NORTH TO SOUTH

Q. When Earth's magnetic field reverses, do the compasses change? Would an electrical appliance run?

A. If there were a sudden overnight flip of the Earth's magnetic field, scouts might be in trouble with their compasses, but most electrical appliances would not be affected. The compass needle is just a little magnet that lines up with the magnetic field lines of Earth; these imaginary lines converge on a spot near Greenland.

In a field reversal, these fields would change, and the "north pole" of the needle would indeed try to line up with the south pole of the terrestrial magnet.

As for appliances, the magnetic field of an electric motor such as the one in a blender or refrigerator compressor is far more powerful than that of the Earth. A motor's magnetic field is typically measured in the hundreds of gauss, the unit of magnetic induction, while the Earth's is a mere half gauss, so a shift in Earth's polarity would be of little moment to a running motor.

ROCKS THAT TELL A STORY

Q. My students tell me that rocks found on Earth show that Earth's magnetic poles have reversed. Is this true?
A. Your students are right.

In the 1960s, an Australian graduate student found rocks from an ancient aboriginal campfire, perhaps 30,000 years old, that were lying in place, magnetized in the opposite direction from the present-day orientation of the earth's magnetic field. The student suggested to his amazed professor that this was evidence of a shift in the magnetic field in such a way that a compass needle would have pointed south 30,000 years ago.

Since then, studies of successive layers of cooled volcanic lava flows and of some sedimentary rocks on the ocean floor have enabled scientists to determine approximately when several reversals of the earth's magnetic field have taken place over the eons. This field of study is called paleomagnetism.

What happens is that heating destroys a rock's original magnetism; upon cooling, it becomes magnetized in the direction of the surrounding magnetic field. This is because groups of atoms in magnetizable material organize themselves in the direction of the field when the material is hot, and are then frozen in place when it cools.

Q. How do scientists distinguish between aftershocks and new earthquakes? I understand that aftershocks can sometimes occur years after the original quake.

A. It is partly a matter of definition, based on the sequence of events and on the magnitude readings and other accumulated data about a given large seismic event. Seismologists define not only aftershocks but also foreshocks.

Foreshock and aftershock are relative terms. Foreshocks are earthquakes that precede larger earthquakes in the same location. Aftershocks are smaller earthquakes that occur in the same general area days to years after a larger event or main shock and in the span of time before the level of seismic activity returns to the normal or background range. (The general area of a shock differs from fault to fault and is defined as being one to two fault-lengths away.)

Aftershocks represent minor readjustments along the portion of a fault that slipped during the main shock. The frequency of these aftershocks decreases with time.

Historically, deep earthquakes are much less likely to be followed by aftershocks than are shallow earthquakes.

TAKING THE MEASURE OF A RIVER

Q. When rivers flood, how are things like flood stage and crest determined? How do they predict when a river will crest?

A. *Flood stage* and *crest* have fairly simple definitions. The flood stage is the height of a river above which damage begins to occur, typically because the river begins to overflow its banks. The crest is the highest level that a river reaches.

Prediction is a more complex matter, involving thousands of readings of river stages and water flows and computer models of likely outcomes. The predictive models are based on experience and must be updated with changes such as each new rainfall. The U.S. Geological Survey monitors flood conditions in cooperation with the National Weather Service.

A river's stage, whether it is flooding or not, is the height of the water surface above a predetermined reference elevation. River stages are continually monitored by the Geological Survey, typically with automatic measuring devices linked to satellite transmitters, so that the data can be obtained even in extreme weather. If the stage of the streambed is known and is subtracted from the water-surface stage, then the result is the depth of water in the stream.

Monitoring devices called stilling wells are sheltered tubes linked to the water with several inlets. They "still" the waters so that momentary surface fluctuations such as waves and surges are smoothed out. Inside, a float goes up and down with the water level, moving a pulley that drives a recording device.

Other devices, called bubblers, measure changes in water level indirectly, by bubbling gas out a tube below the water surface. Changes in the pressure on the gas reflect changes in the depth of the water over the outlet.

Measuring water volume, or discharge, involves physical measurement of the cross-sectional area of the stream and the velocity of the stream. The discharge in cubic feet per second is determined as the product of the area times the velocity.

Velocity is measured by using a current meter with a propeller that is rotated by the action of flowing water. The rotation depends on the velocity of the water passing by. With each complete rotation, an electrical circuit is completed and recorded. A stream is marked off into sections and separate measurements are taken for each segment of width, depth, and average velocity of flow. The total of these measurements is the discharge of the river.

AFTER THE DELUGE

Q. Do scientists suspect the flood described in the Bible was caused by an earthquake like the one in 1999 along the Anatolian fault line?

A. Rising sea levels worldwide following the last Ice Age are a more likely explanation for a great flood about 7,500 years ago, or 5500 B.C., that suddenly inundated areas around the Black Sea. Some scientists suggest it could have been the basis for flood stories in both the Hebrew Bible and the Babylonian epic of Gilgamesh, though others say the accounts came too long after the event to have been inspired by it.

The outlines of such a flood were reconstructed by American scientists from seismic soundings of the seafloor and sediment data gathered by a Russian research ship in 1993, tracing former shorelines. The dating of the flood was based on radiocarbon dates for the transition from freshwater to marine organisms in the cores.

Before the flood, the scientists say, the Bosporus was a narrow spillway carrying freshwater from the Black Sea to the Aegean and Mediterranean. Then there was an abrupt rise in the level of the Black Sea, perhaps a foot a day at the height of the flood, as the flow of water through the Bosporus reversed in a great torrent. In less than a year, the Black Sea became brackish and its level rose as much as 500 feet. Land around it was inundated at a rate of half a mile to a mile a day, leaving about 60,000 square miles of formerly dry land covered by water.

How Murky Water Gets That Way

Q. Gazing upon the clear, aqua-colored water ringing Bermuda, I wondered why southern waters look that way while shallow waters off New Jersey, for example, are murky. What accounts for the difference?
A. Questions of pollution aside, the answer, oceanographers say, is that cold northern waters tend to teem with great concentrations of tiny plants and animals (phytoplankton and zooplankton), while southern waters are (in terms of numbers of individual organisms) empty deserts, with a few species-rich oases, such as coral reefs and waters fed nutrients by river runoff.

Phytoplankton need sunlight, nutrients, and carbon dioxide to thrive. Cold waters are richer in nutrients like oxygen and in carbon dioxide because the water molecules move more slowly than they do in warmer waters, allowing more dissolved gases to stay dissolved. The result is seasonal crops of the phytoplankton, which in turn feed the zooplankton.

This rich soup of microscopic plants and animals and decaying plants makes cold northern waters appear murky. By contrast, a tropical sea has almost nothing suspended in the water to refract light or block the view.

PUTTING THE PRESSURE ON WATER

Q. Is the freezing point of water altered by pressure? Why does deep sea water not freeze?

A. The freezing temperature of water drops about 1 degree for every 75 of the units of pressure called atmospheres, but the change is small compared with the difference even one atmosphere of pressure makes in raising the boiling point. (One atmosphere is equal to 14.69 pounds per square inch.)

Why deep sea water does not freeze depends less on pressure than on two other factors, salinity and heat. The ocean is salt water, which interferes with the freezing process and lowers the freezing point, and at depths of a mile or so, there is considerable heating from inside the earth.

At normal atmospheric pressure, water freezes at 32 degrees. But when water is compressed to about 20,000 atmospheres and cooled, other varieties of ice form, with different molecular arrangements and linkages among hydrogen and oxygen atoms.

The form called Ice II is 12 percent denser, and Ice III is 3 percent denser. The known types go up to Ice X, beyond the infamous "Ice Nine" in Kurt Vonnegut's 1963 novel *Cat's Cradle*.

Ice IX exists, but it does not have the properties Vonnegut gave it, like forming at room temperature and being able to spread quickly. It was discovered after the novel came out, so his version was based on good speculation.

THE TREELESS PLAINS

Q. If trees grow out of rocks with virtually no soil, why were there no trees to speak of on the prairies of the Great Plains?

A. One important reason is that the complex ecology of the prairie environment is formed by a cycle of fires and regrowth of a succession of fire-tolerant plants, and trees do not survive the periodic fires. The other limiting factor is a dry climate, affecting trees more than grassland plants, which hoard water.

The fires sustained the prairies by clearing dead thatch, fertilizing the soil, and destroying trees. Perennial grasses, with their extensive root systems forming an underground sponge, survived both fires and dryness.

In efforts to restore remnants of the tallgrass prairie of the Middle West, ecologists have employed controlled burning to destroy invading trees that suppressed the original vegetation. These fire-adapted species included goatgrass, rattlesnake fern, ebony spleenwort, lady's tresses, and foxtail barley.

Before European settlement, the tallgrass ecosystem covered more than 220,000 square miles, from Canada into Texas and from Nebraska to the Great Lakes. In the dry season of midsummer, most of the fires there were probably caused by lightning. In spring and fall, most were probably set by Indians. The quick new growth was grazed upon by the great herds of buffalo, who left behind dead grasses and stalks that fed the next round of fires.

BENDS IN THE RIVER

Q. Do meandering rivers meander more over time or do they straighten out?

A. There are three basic channel patterns for a river: straight, meandering, or braided. Depending on such factors as terrain, soil, and water volume, different stretches of a river may show different patterns, and channels may shift over time.

There is no fixed rule for the evolution of a river's channel, but one typical progression is from meandering to even more convoluted to straight. What happens in such a case is that the flowing water, which reaches its maximum velocity at the outside of a bend, causes the greatest amount of soil erosion there. The eroded material is deposited as what is called a point bar on the inside of the bend.

The position of the strongest current shifts from side to side, and the meanders migrate and accentuate over time. The migration can be very rapid: Some Mississippi River meanders, for example, move as much as 65 feet a year. As the looping meanders become more pronounced and migrate, getting closer and closer together, they are vulnerable to being cut off during a flood. Eventually the river bypasses a loop, and an oxbow lake is formed as the main course of the river takes the straighter path, past the cutoff lake.

BIG STONES AND SMALL

Q. Farmers often have to clear their fields of big stones. Is this because the stones keep working their way up?
A. Physicists say it is not so much that big stones work their way up as it is that smaller particles work their way down.

Physicists have long known that in a mix of large and small objects, the large ones tend to end up higher while the small ones end up lower. This holds true whether the substance is powder, cornflakes, nuts, or fieldstones, and regardless of the texture, shape, or weight of the particles.

It was not until 1987, however, that scientists at Carnegie Mellon University in Pittsburgh published the results of a simulation study that appears to explain why this happens.

The scientists created a computer model based on particles left to interact with one another completely at random. With each shake of a simulated container, the particles bounce a little, lose energy, and settle down again. As particles bounce upward, gaps open up beneath them, sometimes large gaps, sometimes small, with small gaps much more common.

As the particles fall, a small particle can move down into either a large or a small gap, while a large particle can move down only if many small particles happen to have moved out of the way. That happens, but not often. So over time the small particles tend to head downward and the large particles upward.

WHEN THE EQUINOX COMES EARLY

Q. Our sixth-grade class charted the number of daylight hours based on the times of sunrise and sunset listed in the weather report. We were surprised to find that in New York City on March 17, there were exactly twelve hours of daylight, with sunrise at 6:05 A.M. and sunset at 6:05 P.M. That was three days before March 20, the spring equinox. Shouldn't that be the day with twelve hours of daylight?

A. It would be if this were an airless world with no atmosphere. However, Earth's atmosphere acts like a lens and bends, or refracts, the Sun's light above the horizon. When you see the disk of the Sun resting on the horizon at sunset, you are looking not at the Sun but at a refracted image of the Sun, which is already below the horizon.

The image lingers about three and a half minutes after the actual sunset. For the same reason, the Sun's image appears a few minutes before sunrise.

On average, the atmosphere's lensing effect adds about six or seven minutes to the duration of daylight. Day and night again appear to be equal a few days after the autumnal equinox.

SAVING LIGHT, OR WHAT?

Q. What is the rationale behind daylight saving time? You can't really "save" daylight.

A. The chief practical reason for setting the clock an hour later for a certain period of the year, so that the Sun sets later by the clock, is not to "save" daylight, but to make daylight hours coincide more closely with the time when most people use energy.

A major effect of daylight saving time is that it leaves more hours of natural light at the end of the day, reducing the time between sunset and the typical bedtime.

Energy use and the demand for electricity for lighting homes is directly connected to when people go to bed and when they get up. Bedtime for most people is late evening throughout the year. And when people go to bed, they turn off the lights and TV.

In the average home, 25 percent of all electricity use is for lighting and small appliances, like televisions, VCRs, and stereos.

Studies done in the 1970s by the U.S. Department of Transportation found that the country's electricity usage went down by about 1 percent each day of daylight saving time.

Many people have suggested that the time shift be extended or even kept all year, as has occasionally been done during wars or energy crises. But the savings depends on having people set their bedtimes and rising times by the clock, not the Sun.

CHAPTER 3

The Weather, Rain or Shine

LAYING OUT THE ODDS

Q. When weather forecasters say there is a 50 percent chance of rain tomorrow, what does that mean? And how far ahead can a reliable forecast be made?

A. When the National Weather Service forecasters use the term *probability of precipitation*, they mean that in ten instances of weather patterns similar to the one they are looking at, it rained five times.

The service keeps cumulative records, and the probability forecast is a statistic based on what has happened. For example, if a cold front is expected in mid-July in late afternoon, and it has rained seven out of ten times in such a situation, there is a 70 percent chance of rain. The forecast covers precipitation at any time in a stated period, such as today, tonight, and tomorrow.

As to how far ahead it is practical to forecast, there's a high degree of accuracy up to about three days. But after that, for four to seven days ahead, forecasts are less reliable.

Forecasters can also give a sense of what the weather might be like up to two weeks in advance as a trend—for example, wet or dry and warmer than normal or cooler than normal, based on a pattern that favors hot and dry or a pattern that favors cool and dry.

THE COLOR OF CLOUDS

Q. What determines the color of clouds?

A. A cloud's color depends chiefly on the cloud's relationship to the light of the Sun. In some cases, it depends on the color of the surrounding sky.

If a cloud covers the whole visible area of the sky, the depth of the cloud determines the color of the cloud; the deeper it is, the darker it is. As a ray of sunlight hits  the cloud, the water in the cloud can either reflect the light back toward space or absorb it. Either effect will dilute the strength of the sunlight that gets through.

If the wavelengths of the Sun's visible light, in the familiar red to violet spectrum, are not blocked but scattered equally, the result is white light. So if the Sun is shining and the cloud is not covering the Sun, in general the cloud is white.

Sometimes particles in the atmosphere and the angle of the path the Sun's rays must travel through can result in preferential scattering of wavelengths so that a particular color, like gold, predominates in a cloud. This is the same reason the sky is predominantly blue, because the blue wavelengths are the most scattered.

At other times, when the sky is orange, for example, some of that light is reflected off a cloud, just like a colored light shining on a mirror, so the cloud appears orange.

CLOUD CAPACITY

Q. Clouds seem so light and airy that they couldn't possibly hold as much water as floods down in a storm. How large must a cloud be to hold, say, a gallon of water?

A. It would take a cloud about 60 feet by 60 feet by 60 feet to hold enough water vapor to condense into a gallon of water.

The calculations assume a liquid water content of roughly between ⅓ and 1 gram per cubic meter of water vapor. The amount varies with pressure and temperature; the higher the temperature, the higher the vapor pressure and the more water a cloud can hold. (At a higher altitude the pressure would be lower, and a given volume of cloud would contain less water.)

TRAIL OF THE JET

Q. Why do contrails form?

A. Contrails, or condensation trails, are clouds of ice crystals formed when hot, moist air from aircraft engines mixes with cold, drier air in the upper atmosphere.

Contrails are similar to cirrus clouds, naturally occurring clouds found at altitudes around 20,000 feet.

Portions of the upper atmosphere are too dry to produce condensation, so aircraft flying through them do not leave contrails.

However, significant areas are cold enough and contain enough water vapor for ice particles to grow, provided small particles are already present. It is hard to make ice form from water vapor unless solid surfaces are present, and such surfaces presented by existing ice crystals are rare in the upper atmosphere. Ice can form only if the proportion of water vapor in the cold air is particularly high. In such conditions, the addition of a small amount of water vapor from an aircraft can trigger the formation of a contrail or a cloud. A somewhat similar mechanism is at work on a cold day when your moist, warm, but cloud-free breath meets cold, cloud-free air and a miniature cloud forms.

SOWING THE SKIES

Q. Whatever happened to seeding clouds to make it rain?

A. Cloud-seeding experiments and government programs continue, but in this country at least, they are not regarded as a panacea for drought. The programs are usually subject to strict controls. Some thirty-two states have enacted laws regulating who can practice cloud seeding.

Cloud seeding dates from experiments in 1946 at the General Electric Laboratories in Schenectady, New York, and usually relies on silver iodide as a seeding agent, providing nuclei for the rapid formation of ice crystals that may eventually produce raindrops.

There are some unknowns about effectiveness and safety. For example, seeding at the wrong time and in the wrong place may actually decrease rainfall. Some skeptics worry that there is a possibility of creating severe weather such as hail or floods or of increasing the rainfall in one region at the expense of another. Other scientists say the possibilities of creating hail are remote and that seeding can be suspended if there is danger of flooding.

Cloud seeding has been used for more than twenty-five years in parts of West Texas and has been an important part of water policy in Utah, one of the driest states, since 1973. There have been recent major programs in Florida, Thailand, and Cuba.

MEASURING RAIN

Q. I recently bought a rain gauge, a cylinder with a flared upper rim and a scale in inches on the side. The diameter at the bottom is about two inches and at the top about four inches. Why is it shaped and marked that way?

A. A rain gauge shaped like a megaphone is not the standard shape meteorologists use, but it should work if it has the correct ratios between the cross-sectional area at the top and the part of the gauge that actually collects and measures the water.

If this gauge was a straight-sided cylinder that did not

narrow toward the bottom, it would be very difficult to measure small amounts of precipitation. It would require the bottom inch, for example, to be divided into 100 separate markings for hundredths of an inch, and most rainfall is much less than an inch.

Meteorologists commonly use a gauge that consists of a small straight-sided cylinder within a bigger cylinder. The rain is collected by a funnel at the top of the smaller cylinder, which is only a fraction of the diameter of the bigger cylinder. The inner tube is open at the bottom. The markings on the outer tube reflect a ratio of ten to one, so that an inch of rain would reach a mark ten inches from the bottom.

It's Raining Hamburger Buns

Q. Why are raindrops round? Or are they?
A. Considering the physical forces on precipitation as it falls, drops would probably be more properly described as shaped like a hamburger bun, flat on the bottom but rounded on the top. The shape of the drop as it comes down is ultimately a product of the interaction between the force of gravity acting downward and air resistance acting upward.

In addition, the size of the raindrop matters. The smaller ones tend to be more spherical, while the bigger ones are probably flatter, similar to the hamburger bun. The air resistance they meet on the way down tends to flatten them on the bottom, while they retain the rounded shape on top.

When precipitation is forming in the clouds, rain often starts as snow. Once the snowflakes melt, when they encounter air at a temperature above freezing, they initially have a more spherical appearance, and they would keep that shape if there were no air.

When they encounter air resistance, they behave the way something that is spherical but pliable, like a balloon, would act when pushed upward by the palm of the hand.

BAD TIMING

Q. Why does there seem to be a thunderstorm in the Rocky Mountains every time I try to have a summer picnic?

A. Summertime conditions in the Rockies constitute a recipe for thunderstorms. The ingredients are strong sunshine coming at a direct angle to warm the land below, sharp differences in temperature between air warmed by the mountains and cooler air over adjacent plains, and enough water in the air.

Thunderstorms owe their existence to thermals, or localized pockets of warm air that rise, and the topography of the Rockies is predisposed to create intense thermals. The sun heats a mountain chain averaging 8,000 feet in elevation much more than it heats the low-lying plains, and the mountains in turn heat the air. The magnitude of the updraft is  directly proportional to the intensity of the heating, and the summer sun is the hottest.

When the resulting warm bubble of air accelerates upward because of its pressure difference, it can cool to the point that the water vapor within it condenses, and if there is enough, precipitation occurs.

It so happens that summer airstream patterns bring in plenty of moist air from both the Gulf of Mexico and the Gulf of California; at other times of the year, there is more dry air from the continent. The outcome is a cloud, rain, and a ruined picnic.

HOT AIR, COLD PEAKS

Q. If hot air rises, why do we get snowcapped mountains?

A. Hot air undoubtedly rises, because it is less dense than the cold air next to it. However, as the air moves up, it expands and gets colder. It is the same expansion-and-cooling principle that makes a refrigerator work, because the molecules are moving away from one another and slowing down as the pressure steadily decreases.

The pressure decreases because as the hot air rises, there is less and less air above that weighs down on it. Therefore, the air at higher altitudes is almost always cold, no matter how hot it was to begin with, and by the time it gets to the top of a high mountain, it is cold enough to allow for snowcapped peaks.

BENDING SOUND

Q. Does weather, especially humidity, affect how far a train whistle carries?

A. Humidity is one of several atmospheric factors affecting the propagation of sound. Heavy, muggy air actually carries sound better. Bone-dry air greatly attenuates sound.

But humidity is not as prominent a factor in affecting low-frequency sound as the bending of sound rays (or sound waves going in the same direction) by wind and the temperature profile of the air.

A steady wind blowing toward you from the sound source increases the sound, and one blowing away from you decreases it. The rays that would otherwise go out to space are bent down toward you if the wind is toward you, up if the wind is away. With some loud sounds, like artillery fire, people very close hear the sound, then there is a dead zone, and then people farther away hear it again.

This lensing effect leads to dangerous situations at unguarded railroad crossings because the whistle of a high-speed train may not be heard until it is quite close to the road.

The other lensing effect occurs in a temperature inversion, the weather pattern that traps smog in California. In a normal atmosphere, as you go up in altitude, the temperature drops. In a temperature inversion, as you go up, the temperature rises, rather than falling, and this type of temperature profile bends rays down toward the receiver.

VOLCANOES AND EL NIÑO

Q. Is it possible that warming in the Pacific Ocean is caused by undersea volcanic action?

A. Some scientists think it is possible that volcanic activity might be at least partly responsible for the periodic weather phenomenon called El Niño. This theory was inspired in part by the discovery in 1993 of the largest known cluster of volcanoes, covering an area the size of New York State. The area in question is 600 miles northwest of Easter Island, and it was found to have 1,133 seamounts and volcanoes, far more than any other area then known on land or under the ocean.

While many researchers dismiss the theory, and even proponents concede that there is no proof of cause and effect, some scientists say the idea is plausible enough to merit more research.

When the cluster of volcanoes was discovered, scientists said they were interested in determining whether periods of extreme volcanic activity underwater, like the simultaneous eruption of a dozen or more volcanoes, could be enough to set off El Niño, which alters global weather patterns every two to seven years.

When El Niño is active, a giant high pressure system centered near Easter Island drops slightly in pressure and touches off a change in Pacific currents, bringing warmer water to certain areas and colder water to others. In 1995, scientists reported that a spate of undersea quakes and lava eruptions had preceded the onset of the El Niño cycle then in effect, leading some scientists to tentatively link the seismic activity to El Niño.

INVISIBLE ICE

Q. Is black ice really black? What is it, exactly?
A. What weather reports call black ice is not black at

all; it just looks that way because it is transparent and whatever color the road's surface is shows through.

Whether ice on the road is visible depends on how it is formed. Rime ice, which is visible, forms when droplets of freezing drizzle adhere to the roadway with air trapped inside because the drops do not splatter when they hit.

Black ice, which is not eas-ily seen, is a coating of thin ice that is often the result of thawing and refreezing road-side snowbanks. It may also come from freezing drizzle or rain that splatters when it hits the road, so that no air is trapped inside.

In daylight, black ice may look like a dull patch of road surface that is darker than the rest of the highway. It is most treacherous at night, when it is difficult or impossible to spot. Unfortunately, it is also most com-mon at night, because that is when temperatures drop, traffic flow eases, and melted snow refreezes.

Black ice should be suspected, and driving speeds should be reduced, when temperatures are near or below freezing, especially in fall and early winter, when there is no obvious precipitation on the road.

CHAPTER 4

Food, Glorious Food

The Nature of the Yeast

Q. What is the stuff that comes in yeast packages? Why doesn't it die in the sealed package?

A. Yeasts are simple one-celled plants, members of the large Ascomycetes family of funguses. The powder or cake sold for baking is billions of cells of one of 160 or so useful funguses that are called yeast. Commercial baker's yeast is a pure, selected strain of *Saccharomyces cerevisiae*. It is a "tame" yeast, raised in factories on a solution of aerated molasses, as opposed to wild yeast strains, which are almost ubiquitous in soil and on plants.

You can catch your own wild yeast in the form of sourdough starter, leaving food like milk out in the open to capture whatever stray yeasts are wandering by (mostly *Saccharomyces exiguus*) and then saving a bit of dough from each batch of bread to start the next. Commercial yeasts are pedigreed and consistent from batch to batch.

Bread yeast ferments sugar to produce alcohol, carbon dioxide, and heat. The carbon dioxide gas bubbles make the bread rise, and the alcohol evaporates. Edible yeasts, which also make beer and wine, synthesize protein, vitamins, and enzymes, and dried yeast is highly nutritious.

In 1876, Louis Pasteur demonstrated that yeast, especially bread yeast, can shift its metabolism from fermentation to oxidation in the presence of air, a shift called the Pasteur effect; oxidation is used in the commercial production of yeast. When yeast is sealed in a package and kept cool, it shuts down most of its metabolic activity, waiting for warmth and food to multiply and raise bread.

HOLES IN THE CHEESE

Q. What makes the holes in Swiss cheese?
A. The holes are made by sizable bubbles of gas produced by a special kind of bacterium in the cheese-ripening process.

There are three basic steps in producing most kinds of cheese: making curds from milk, concentrating the curds, and ripening the curds.

First, hungry bacteria, pure cultures of streptococci and lactobacilli, feed on milk sugar, or lactose, and their digestive processes make lactic acid. The acidic environment is favorable to the action of a curdling agent like ren-

net, a material found in cows' stomachs. The milk separates into the familiar curds and whey.

Then the solid curds are drained of most of the liquid whey and cooked, pressed, and salted to remove the rest.

The final process, ripening, the breakdown of the curd into simpler, tastier molecules, is left to the enzymes produced by a number of microbes, including the molds that produce the veins in blue cheese.

However, it is evenly dispersed starter bacteria that finish off Swiss cheeses, which are ripened from within by the starter bacteria. An additional strain of bacterium, *Propionibacter shermanii,* is included in the starter. It lives on the lactic acid excreted by the other bacteria and gives off prodigious amounts of carbon dioxide gas, which collects in large pockets, or eyes, in the curd. The bacteria also give off propionic acid, which contributes to the flavor.

The characteristic holes fail to develop properly when the casein (the bundles of protein molecules that clump into curds) has been preheated in pasteurization. In the United States, regulations require that most cheeses that are aged less than 60 days at temperatures lower than 35 degrees Fahrenheit be made from pasteurized milk. Swiss cheese is one of the few cheeses exempt from this regulation.

FORMING A SKIN

Q. What is the skin that forms on the surface of milk when it is heated?

A. It is a complex of casein, or milk protein, and calcium, and results from evaporation of water at the surface of the heated liquid and the concentration and coagulation of protein there. Skimming the skin off removes significant amounts of nutrients. Skin formation can be minimized by covering the pan or whipping up a little foam; either slows down evaporation.

BUBBLING AWAY THE BOOZE

Q. Is alcohol removed from wine or liquor when it is used for cooking?

A. Wine that is simmered for a long time in a stew would probably retain only a small fraction of its alcohol, but other dishes, like briefly flambéed crepes Grand Marnier or a soup with sherry added at the last minute, would probably still have enough to be of concern to those who want to avoid it altogether.

A study first reported in 1990 found evidence that alcoholic beverages retain from 5 percent to as much as 85 percent of alcohol after cooking.

In that study, six recipes using wine or spirits were prepared according to directions and analyzed. The study found that the amount of alcohol left behind varied by the type of heat applied (simmering, baking, or flaming), the source of the alcohol (wine or spirits), the length of cooking, and the treatment of the food after preparation and storage.

A sauce made with Burgundy accompanying pot roast retained 5 percent of its alcohol after it had simmered for two and a half hours. Another wine sauce accompanying a chicken dish retained 40 percent of its alcohol. It simmered only ten minutes after the wine was added.

Forty-five percent of the alcohol in sherry used to make scalloped oysters remained after baking for twenty-five minutes. Brandy that was heated and quickly flamed before being served over cherries jubilee retained 75 percent of its alcohol. Two tablespoons of a Grand Marnier sauce, cooked slightly, retained 85 percent of the alcohol. The liqueur was added to a hot liquid sauce after it had been removed from the heat.

A slice of brandy alexander pie, made with brandy and crème de cacao, was not heated. After it sat uncovered for sixteen hours in a refrigerator, the researchers found, it had lost 25 percent of its alcohol, probably through evaporation.

In other words, there can be enough alcohol left in food to be of particular concern to recovering alcoholics, pregnant women, and children old enough to eat the table food served to the rest of the family.

However, as food writers pointed out when the study was first reported in 1990, it seems likely that further heating of the cherries jubilee and cooking of the Grand Marnier sauce would have further reduced the alcohol content.

Gauging a Pepper's Heat

Q. How is the heat of hot peppers measured?
A. The best-known method relies on the amount of a pepper extract mixed with water that a panel of tasters need to sense the heat. Invented in 1912 by Wilbur L. Scoville, a pharmacologist, the rating is given in Scoville units.

The sample is dried and mixed with alcohol; then

sugar water is added. The tasters are given shot glasses holding increasingly more dilute versions of the concoction and are asked to determine the dilution at which the pepper no longer burns the mouth.

A mild salsa might rate a puny 400 units on the scale; a jalapeño pepper might be 1,500 to 4,500, while a much hotter habanero pepper could hit 150,000.

The method relies on the subjective human palate, and scientists have increasingly turned instead to measuring the amount of capsaicin in the pepper. Capsaicin, a substance that gives the sensation of heat when it hits nerve receptors, is analyzed in terms of parts per million using computerized devices.

Substances other than capsaicin may be involved in heating up the mouth, and both ratings can vary widely from crop to crop and even from sample to sample from the same crop.

If eating a highly rated pepper causes distress, drink milk or eat yogurt; casein will help neutralize the capsaicin, while water usually does not provide relief.

PERFECT BANANAS

Q. How does temperature affect ripening bananas?
A. Temperature changes can either delay or hasten the ripening of bananas. The banana is a tropical fruit, adapted to ripen most quickly at a certain stage of its

development and at a particular temperature and humidity. Bananas continue to ripen after they are harvested, with more and more of their starches converted into sugars by the action of enzymes. When harvested, bananas contain about 20 percent starch and only 1 percent sugar. By the time the fruit is ripe, the proportions are reversed.

Bananas also release comparatively large quantities of ethylene gas to help themselves ripen; the gas will even ripen other fruit put in a bag with a ripening banana.

Bananas are usually harvested while still green, cutting off the supply of nutrients at the stem, and then shipped at a temperature low enough to slow the action of the enzymes of ripening. Later, the bananas are brought back up to a temperature and humidity that let the enzymes become active again. Too high a temperature destroys enzymes, and too low a temperature can break down the cell walls of the fruit so the contents mix and the fruit oxidizes, browns, and softens abnormally.

The optimum temperature and humidity conditions for ripening are 62 to 68 degrees Fahrenheit and 90 to 95 percent relative humidity. Storage temperatures should be 53 to 56 degrees Fahrenheit.

THE TELLTALE FISH

Q. Why does fish start smelling awful in the garbage pail so much sooner than meat does?

A. Fish contains quite a bit of a substance that gives rise to two chemicals that make the rotten-fish smell, while red meat has hardly any. The chemicals are trimethy-

lamine and dimethylamine, both derived from trimethylamine oxide, which is particularly common in the flesh
of fin fish, especially cold-water, surface-dwelling ocean
fish like the cod. Ocean fish start to smell faster than,
say, catfish.

When the fish is killed and exposed to the action of
air, its own enzymes, and bacteria, the trimethylamine
oxide undergoes a chemical process called carboxyla-
tion. The products are amines, which are volatile and
quite odoriferous.

Fish dealers test for the amount of amines to determine how fresh fish is. A consumer's nose knows, too,
but those who do not eat fish find it more objectionable
than others.

A TOMATO OF A DIFFERENT COLOR

Q. I see yellow tomatoes, among other exotic types, at
vegetable stores. Do they differ in nutritional value from
red tomatoes?
A. Like all vegetables and fruits, tomatoes differ somewhat in nutritional value from variety to variety, and
even from season to season. For example, the U.S.
Department of Agriculture lists different average nutrient contents for tomatoes harvested from November to
May and for those from June to October.

Many food values are comparable for red and yellow
tomatoes, or are a trade-off. For example, the amino
acids are found in similar smidgens. The precursor of
vitamin A called beta-carotene is found in valuable
amounts in red and yellow tomatoes.

Both colors have vitamin C, though a red tomato has about three times as much. Minerals are similar, though a yellow tomato is higher in sodium. Yellow tomatoes have more niacin and folate; red tomatoes have more vitamin B_6 and pantothenic acid, and so on.

GINSENG PEDIGREES

Q. What is the difference between American ginseng and Asiatic ginseng? What about Siberian ginseng?

A. Asiatic ginseng, *Panax schinseng,* and the American wild ginseng, *Panax quinquefolius,* are closely related and similar in both appearance and effects. Siberian ginseng includes species of a different family, the Eleutherococcus family, which is not real ginseng.

There is not a big medicinal difference between the two "real" ginsengs. Each has a slightly different composition of ginsenosides, active ingredients that can be either stimulants or the opposite; American ginseng is slightly less stimulating than Asiatic ginseng.

The principal use of ginseng is to make people feel better, and there is evidence that it works, especially in older people. Its reputation in ancient Chinese medicine as an aphrodisiac is not scientifically proven.

Prepared ginseng root should be bought from a reputable source. Side effects can result from adulterants used to extend an expensive product. Makers offer different concentrations, from about 4 to 7 percent, so follow the dosage on the label.

Anatomy of a Bagel

Q. How many calories are in a plain, sesame, or poppy seed bagel from a New York coffee shop? What are the ingredients and nutritional value?

A. Let us assume that you get the biggest plain, enriched bagel analyzed by the U.S. Department of Agriculture, 4½ inches in diameter, weighing 110 grams, about 3.8 ounces.

The ingredients—flour, water, salt, yeast, and malt, but no sugar, if it is a classic bagel—are boiled and then baked. They add up to 302.5 calories, the USDA says.

On a standard nutrition facts label, the bagel would boast 1.76 grams of fat, no cholesterol, 587.4 milligrams of sodium, 111.1 milligrams of potassium, 58.74 grams of carbohydrate, and 11.55 grams of protein. Vitamins and minerals include a significant amount of folate, 96.8 micrograms, from the enriched flour, but most others are present in trace amounts. A bagel preserved with calcium propionate has more calcium than one without it: 81.4 milligrams, compared with 19.8 milligrams.

Oddly, the USDA does not differentiate among plain, onion, poppy seed, and sesame bagels. Poppy seed, which the department considers a spice, not a food, would probably not add enough calories to make a weight watcher feel guilty. There are only about 15 calories in a teaspoonful, fewer calories than would be in a spoon of sugar. Sesame seeds have perhaps 26 calories in a teaspoonful, figured at a sixth of an ounce, by volume.

THE PHYSICS OF NOODLES

Q. How are "instant" cereals, rice, and other foods processed to diminish cooking time?

A. It is really a physics problem involving the interaction of water with a carbohydrate, and the process differs from food to food.

When you have a solid food that you want to be able to cook faster, such as rice, you hydrate it, then dehydrate it under controlled conditions, adjusting the chemistry and physics of the particle to maximize its ability to rehydrate. Then when you rehydrate it to cook it, it cooks much faster.

For many foods, the processing involves increasing overall surface area, so the water does not take so long to migrate through the food. For tea and coffee, it involves agglomerization, building up the particle size so the surface area increases dramatically, which makes it easier for water to be absorbed.

For some products, sugar or another hydrophilic (water-loving) additive is mixed in to speed absorption. For noodles or dried vegetables for instant soup, the process may involve quick drying in a vacuum, which puffs up the food so it will rehydrate more quickly.

CHAPTER 5

Flora! Flora! Flora!

STAYING VERTICAL

Q. How do trees on a hillside sense the vertical so they can grow straight up?

A. Trees grow vertically because of two things: gravity and light.

Geotropism, in which microscopic particles in plant cells react to gravity, tends to make the roots grow straight down, which means that the stems grow straight up.

Phototropism tends to make plants grow vertically as well, following the direction from which light comes. Phototropism was originally called heliotropism, or bending toward the sun, until scientists found out that plants would bend toward light in general, not just sunlight. The phenomenon was studied by Charles Darwin and his son Francis, who recognized that the bending started just below the tip. But they did not discover the

mechanism that caused the bending.

Subsequent researchers found that a class of plant hormone called auxins can regulate the growth of plant cells, interacting with other plant substances to direct and control the plant's final shape, both above and below the ground.

In a growing tree, auxins, produced at the growth tip, promote the elongation of plant cells. Auxins are present in greater concentrations on the darker side of the plant shaft, so those cells grow longer than the cells exposed to light. Thus, the tree bends toward the light.

IT'S EASY BEING EVERGREEN

Q. Why do evergreens keep their leaves?

A. Evergreen trees do not keep their leaves indefinitely, but rather may grow new ones before the old ones fall, or keep some and drop others over a period of time.

As any southern gardener can tell you, some broad-leaf evergreens, like *Magnolia grandiflora,* drop their heavy leaves in a staggered cycle over the course of the year. Some, like the live oak, may show a heavy leaf loss in early spring, but retain a few old leaves while the new ones develop.

There is also seasonal needle loss with evergreen species of conifers, like pine, spruce, and fir trees. The leaf drop is a normal process, not a disease. The older, inner needles senesce, or grow old, turning yellow or brown, and drop from the tree after one to several years, depending on the species. By November of most years, for example, white pines may have only a year's worth of needles attached to the tree.

Austrian and Scotch pines usually retain their needles for three years. Spruce and fir needles also yellow and drop with age, but these trees retain their needles for several years, so needle drop is often not noticeable. Arborvitae and some cedar needles usually turn brown rather than yellow with age.

These general patterns of needle drop may vary from tree to tree and year to year; adverse conditions in the summer and fall may lead to a more pronounced needle drop.

UNABLE TO LET GO

Q. Why do some trees hold on to their leaves longer than others?

A. The timing of leaf loss varies with species, site, and season. Day length and temperature are the two triggers for color change and leaf loss. The timing is usually species specific but is also related to site conditions.

For example, a fairly dry midsummer to fall in the Northeast can mean that some trees drop leaves early, before they turn, in a reaction to drought stress; leaves may also die on the tree but hang on until much later.

Species variations are also important. Norway maples normally have green and fully functional leaves that keep on photosynthesizing until two or three weeks after leaves of sugar maples have turned. If both are on a cramped site, Norway maples, with extra weeks of energy storage, may outgrow and outlive sugar maples.

Oaks keep their leaves much longer than many other species because a layer of cells that forms where the leaf stem is attached, called the abscission layer, does not form a complete barrier. In beech trees, which are in the same family, an incomplete layer is seen in younger trees, but mature beeches, twenty-five to thirty years old, form a complete layer.

There are also sex differences; for example, leaves of female ginkgo trees usually color and drop earlier than those of males. And trees near streetlights may be affected by the longer light exposure and keep their leaves longer.

TREES' PARTNERS

Q. I heard there was a fungus that helps trees absorb water. What is it?

A. What you heard about is mycorrhizae, complexes formed by beneficial fungi and plant roots. There are many different ones, and particular plant species form relationships with particular species of fungi.

There are three basic types, depending on where the complex is formed: endomycorrhizae, inside the root hair; ectomycorrhizae, outside the root hair, and endo-ectomycorrhizae, both inside and outside.

The fungus and the plant have a symbiotic relationship, meaning they help each other to thrive. Fungal cells take in some nutrients and water from the soil more easily than plant cells can, while the plant gives the fungus other resources.

Mycorrhizae have been known for many years, but only recently have plant scientists been able to isolate the strains for certain plants and reproduce them en masse for application.

Commercial mycorrhizae preparations are available for many conifers and broad-leaved trees and for certain annuals and turf. The association forms only in the fall, and the fungus normally needs to be injected into the soil.

DOOMED BAMBOO

Q. Is it true that all bamboo in the world dies off at the same time every hundred years or so?

A. No, but certain species among the hundreds or thousands of known bamboos die off all at once after their single episode of flowering. That can create problems for gardeners who cultivate a particular ornamental type.

More seriously, a mass die-off of a bamboo species in the phenomenon called gregarious flowering can endanger a species, like the giant panda, that finds its usual food source, *Gelidocalamus fangianus*, gone from a wide area.

However, flowering bamboos (and not all species are known to flower) vary widely in their cycles; the fixed periods range from a year to a few years or even more than a century. Not all species are so weakened by flowering that they die.

Gregarious flowering usually occurs when all plants from what bamboo enthusiasts call a clone (that is, offspring from a single parent, which has been repeatedly divided and distributed) flower at about the same time.

There has been extensive research on how to keep a stand of bamboo that starts to flower from dying because bamboo grown from seed may differ significantly from the parent plant. Techniques include cutting back flowering culms (or canes), fertilizing generously, and propagating by planting a section of a culm with several nodes or joints.

LOVE THAT SALT

Q. Are there plants that can survive if watered with salt water?

A. Yes, there are such plants, called halophytes. Some are natural species that grow well when the water they get has a high salt content. Others are species that scientists are tailoring for the purpose by selective breeding or crossbreeding for salt-tolerance genes.

Halophytes include grasses, shrubs, and trees. Some are edible crops suitable for areas with salty soil or irrigation water, and others are ground-cover varieties that are useful in wetland restoration.

One edible halophyte is salicornia, which produces an oil high in polyunsaturates. Another crop plant is atriplex, which yields animal feed similar to alfalfa. Still another is tecticornia, which grows in coastal mudflats and has seeds that the Australian aborigines grind into a flour.

Seawater contains 35,000 to 38,000 parts of salt per million, and researchers believe most conventional crops could not be bred to tolerate levels that high.

However, brackish water with salt levels of 5,000 to 7,000 parts per million has been successfully used in drip irrigation for some crops. Drip irrigation provides a constant trickle directly to root areas, keeping them moist but adding a small total amount of salt to the soil.

FROM SILK TO KERNEL

Q. Is it true that each strand of silk on an ear of corn represents a kernel of corn?

A. Yes, but only if pollen falls on the silk. Otherwise, a kernel does not develop. A corn plant produces corn silk surrounding each ear about two months after the plant emerges from the ground. The plant matures after about four months.

Tassels, the male part of the plant, emerge at its apex and shed pollen for a week or two, fertilizing the individual silk strands below. The silk strands grow a little more than an inch every day and continue to grow until they are fertilized.

Pollination occurs when the falling or wind-borne pollen grains are caught by these new moist silk strands. A captured pollen grain takes about twenty-four hours to move down the silk to the ovule, where fertilization occurs. The ovule then develops into a kernel.

Generally, two to three days are required for all silk strands on an ear to be exposed and pollinated. Ovules that are not fertilized will not produce kernels and will eventually degenerate. Stressful conditions, including lack of moisture, tend to dry the silk strands and pollen grains, usually resulting in a nubbin, or an ear with a barren tip.

Waiting for Roses

Q. Can I grow a rose from seed? How long would it take?

A. With patience, a rose can be grown from a seed, and indeed all the Peace roses of today are descended from a plant grown in 1939 from a single seed nurtured by Francis Meilland, the hybridizer. But a random seed from a Peace rose or any hybrid would be unlikely to produce a plant like the one it came from.

As for the time it would take, a wild, unhybridized rose like *Rosa rugosa* would be relatively quick and easy to grow from seed, with little preparation; in the early fall, remove the seeds from the ripe rose hips, the fruit of the rose, and plant them.

For many other roses, however, the seeds must be refrigerated for about four weeks at about 40 degrees,

for what gardeners call stratification, a cold period, and again for about four months, at a somewhat higher temperature, for germination, with a light on in the refrigerator. They can be nurtured in moist peat moss.

Before sprouting, seeds

can be tested for viability by seeing if they float in water. Floaters are hollow and unlikely to germinate.

The sprouts from the refrigerated seeds can be planted, preferably in sterile soil, perhaps with a cutoff plastic bottle as a greenhouse. If they survive, they may produce miniature roses in just a couple of months, previews of the color of mature roses, but not their shape. The plants and their roses should reach full size in about three years.

The Dreaded Rhubarb

Q. Are rhubarb leaves toxic?
A. While not poisonous, large quantities of rhubarb, stems or leaves, offer possible risks, and the stems may be as bad as the leaves for susceptible people. To complicate things, some rhubarb chemicals can have positive or negative effects.

In cell cultures using an extract of the entire plant, studies found some antimutagenic effects, depending on whether water or organic solvents were used.

Rhubarb also contains some chemicals called sennosides that are used as laxatives. Very large quantities can cause severe dehydration. The laxative effect may be where the idea of a poison arose, but the laxatives are in the stems, too.

When one sennoside, called emodin, is metabolized by the liver, one of the products, under alkaline conditions, has weak mutagenic effects, but in tests, emodin halted cancer cell growth in the test tube.

Rhubarb also contains oxalic acid, which is somewhat toxic. If there is a difference in the concentration in leaves and other parts, it is less significant than the fact that rhubarb has a high oxalic acid content overall. Oxalic acid scavenges calcium and magnesium, and a buildup may be risky for people genetically predisposed to kidney stones.

PLANTS AS INSECT REPELLENTS

Q. Can you use the Osage orange to repel insects?

A. It may be that the rich orange smell of the fruit, which folklore has long rumored to be an insect repellent, scares them off, or it may be that the latex-rich leaves gum up the mouth parts of insects, eventually discouraging others. You could try leaving the fruit around your house, or experiment with planting it or some of the other fragrant plants that have been reported to keep insects away.

The Osage orange, *Maclura pomifera*, a spiny shrub commonly used as a living fence in the American Southwest, is just one of the plants containing chemicals that may (or may not) help repel some or all insects. There is a wealth of lore about them, and many anecdotal reports of success, though relatively little hard data. Like the Osage orange, many of these plants are also sources of perfume fragrances.

Depending on climate and space, backyard experimenters could include some of the following candidates: marigolds, chrysanthemums, basil (for mosquitoes), rosemary (for moths), patchouli, vetiver, citronella, peppermint, spearmint, and pennyroyal. The tree spices cinnamon and cloves are also reputed to be moth repellents.

Gaining a Toehold

Q. How does mistletoe nourish itself on oak trees?

A. Members of the Loranthaceae family like the American mistletoe, *Phoradendron flavescens,* take food from the sap of woody plants through special roots called haustoria. Mistletoe also grows on other deciduous trees, including the sweet gum and red maple.

Man-Eating Plants

Q. Are there any carnivorous plants that are harmful to humans?

A. No, because even the largest carnivorous plants are only big enough to digest something the size of a frog, and the digestive enzymes they secrete are comparatively weak.

Very rarely, a rat or bird has been found drowned in the liquid contained in the bottlelike appendages that hang from the vines of *Nepenthes rajah,* a pitcher plant from the tropical highlands of Borneo. The bottles can be up to 14 inches long, with mouths as big as 7 inches.

However, frogs are more typical Nepenthes prey, and birds or rodents that get caught are probably sick animals too weak to fight their way out of the trap. Most carnivorous plants are completely harmless to something larger than a small insect.

Botanists say that the many kinds of carnivorous plants evolved to get nutrients, especially nitrogen, that are lacking in the soil or water where they grow. Some live with their traps submerged in water, eating very small prey up to the size of mosquito larvae and fish fry.

Some land-dwelling carnivores consume gnats, flies, and moths; others eat ants and other crawling insects. The most familiar carnivorous plants, Venus flytraps, will capture and digest any insect or spider that happens their way.

CHAPTER 6

All Bugs Great and Small

BUGS ON THE WINDSHIELD

Q. When I was a child, clouds of bugs used to hit the windshield on vacation drives. What were they?

A. Depending on the part of the country, time of year, and time of day, they could have been anything from flies to moths to dragonflies. Especially in the South, the most likely culprit is flies, and fly populations are highly cyclical.

In coastal Florida, Alabama, Mississippi, and Louisiana, lovebugs—the bibionid fly species *Plecia nearctica*—form clouds twice a year during their spring and fall mating seasons. The problem comes and goes, with some years being much worse than others.

Lovebugs form linked pairs during their nuptial flights. The windshield smashing may occur in mating season.

In Florida, it was discovered that part of the problem

was that the flies were breeding and feeding in decaying grass clippings along the highways, left there by state mowing projects. The state changed the mowing schedule.

A Backyard Census

Q. How many insects and other arthropods might you find in a northeastern backyard?
A. The number of individuals and species would depend on many variables, like soil condition and chemical use, but if the first few inches of soil were included, the totals might be impressive.

One survey of backyard insects in the 1930s collected 1,402 species, including 467 butterflies and moths, 258 flies, 259 beetles, and 167 wasps, bees, and ants, from a 75-by-200-foot garden in a northeastern New Jersey suburb. The survey did not include arachnids like spiders and mites.

A suburban lawn is a monoculture, a single-crop field, and since, in general, biodiversity increases with complexity, the number of large, easily visible arthropod species on the surface of the soil or in the grass itself would be fairly low. One estimate is 25 to 50, with the number greatly reduced if pesticides had been used.

However, if the fauna of the litter layer and the top few inches of the soil is included, a survey could find huge numbers of individuals of springtails and mites, as many as tens of thousands per square yard. Up to 50 species of springtails per square yard have been found in woodland soil.

THE SWARMS OF THE ANNOYING

Q. Why do gnats and mosquitoes and other annoying insects get together in swarms?

A. Many insect species swarm, both the beneficent (like honeybees) and the irritating (like mosquitoes). In many cases, springtime swarms consist of groups of males all looking for suitable mates. Pheromone signals probably keep the group together. As insect mating, egg laying, and hatching tend to take place in synchronized waves, it is not surprising that a generation emerges with one goal.

Formation of mosquito swarms, which vary from species to species, can depend on variables like available space, lighting conditions, and winds. Mosquito swarms often focus on a defined point like a bush or hedge, where the air is undisturbed, and there are even reports of swarms that form over the head of a person and follow him around as he slowly walks.

Other large insect groups, all moving in the same general direction, may be migratory insects, like monarch butterflies, or social insects setting up new colonies.

The next time a cloud of mosquitoes buzzes around your head, count yourself lucky. The desert locust, *Schistocerca gregaria*, can blacken the sky for miles around. A swarm in Kenya in 1954 was estimated to cover more than 77 square miles.

Q. Do aphids milk ants or vice versa? In any event, how can I rid my houseplants of the aphids?

A. Several ant species "herd" the small pear-shaped insects called aphids and "milk" them for a sweet fluid that they excrete after feeding on sap from roots or leaves. They are sometimes called dairying ants. Ants that "milk" root-eating aphids often dig them chambers to serve as "barns."

The ants protect their aphids, carry them from plant to plant, and gently stroke them with their antennae to get them to release droplets of the liquid, called honeydew. Some species of yellow ants called Acanthomyops feed almost exclusively on honeydew harvested from roots by aphids.

Aphids harm houseplants and some crops by sapping them of their fluid and sometimes by coating the leaves with so much honeydew that mold grows, inhibiting photosynthesis.

Agricultural extension agents recommend some steps to counter aphids on houseplants without using commercial insecticides. First, isolate new plants until it is obvious that they are not infested, or remove the infested leaves. If an infestation is advanced, try swabbing or spraying the leaves with rubbing alcohol or soapy water, repeating at intervals of a few days to get rid of eggs and new hatchlings; rinse off the soap. Or buy ladybugs, which feed on mites, not plants.

Lines of Communication

Q. Watching ants traveling in a line, I noticed they meet head-on for a moment, then continue on their way. Is this some type of communication?

A. If there was any foreleg contact or waving about of antennae by the ants, some communication may have been involved. But the wide range of communication methods among ants is weighted heavily toward chemical signals, and visual signals have not been solidly documented.

Ant communication, which has been extensively but not exhaustively studied, includes tappings, chirplike grating noises, strokings, graspings, nudgings, tastings, and streakings of chemicals, as well as activities like the exchange of special eggs and anal secretions.

In a typical worker ant, more than ten organs have been implicated in producing just one class of chemical signal. However, a typical ant colony operates with only about ten to twenty signals, most of them chemical.

If a fire ant is laying a food trail using such signals and meets another worker, the homeward-bound worker may turn to face her, rushing into her before continuing on or even pausing briefly to shake her, but researchers think this behavior may serve only to enhance the effects of the chemical being used.

Calling In Reinforcements

Q. Are there really caterpillars that scream to protect themselves? If so, what types? Can we hear the sound they make?

A. While they do not exactly scream, some caterpillars have noise-making organs, and the sound they produce is used to call in troops of ants to protect them from predatory wasps. The noise has an average frequency of 1,877 hertz, which would be audible to human ears if it were not so very faint.

Some noisy butterfly larvae species belong to the Riodinid family and live in the tropical rain forests of the New World. Members of the Lycinid family, which has a worldwide range, also make sounds, for similar purposes.

The caterpillars' sound-producing organs and their relationship with ants were first reported by Dr. Philip J. DeVries in research published in 1990. There are actually two sets of organs: just behind the head is a pair of grooved rods, and rows of projections resembling guitar picks protrude from the top of the head. Dr. DeVries described the noise as being like the sound of a comb running over the edge of a table. The caterpillars reward the ants with amino acid secretions that the ants feed on.

BUTTERFLIES OVER BROADWAY

Q. In mid-September, I saw many monarch butterflies headed south in midtown Manhattan. Is this unusual? Do they congregate around New York City somewhere?

A. That is the right time of year for the eastern monarchs' migration, and they do congregate in Cape May, New Jersey, by the thousands.

The monarchs seen in and around New York City are flying through on the way to their wintering sites in the mountains of central Mexico. The butterflies seek out microclimates congenial to an overnight rest and are off again in the morning.

The migration happens every year, but is much more noticeable in some years because insect populations are highly variable. Factors such as wind patterns may influence the exact route. However, each year there are reports of sightings in Manhattan, even of big groups flying by office windows thirty floors up.

The eastern population of monarchs overwinters in the Transvolcanic Range in Mexico, on the border of the states of Michoacán and Mexico. The site, 10,000 feet above sea level, was discovered in 1975. The western population winters on the coast of California.

FAT FOR THE FLIGHT

Q. With its small body mass, how does a monarch butterfly obtain enough energy to migrate 1,800 miles?
A. The energy is gathered from nectar, and the butterflies that make the trip, those born in the early fall, are able to convert nectar into fat.

Those born in September know to fill up on nectar. Their abdomens get really large. Unlike the other generations, these monarchs have a little area of fatty tissue where the sugar of the nectar is converted into fat. They can live off this cushion in winter and need only water to rehydrate their bodies.

Monarchs born in September or late August live seven, eight, or sometimes nine months. Their children, grandchildren, and great-grandchildren live just a month. By the time you get to the great-great-grandchildren, you're back at September.

DRAGONFLY DIPPING

Q. I have seen dragonflies that appear to be dipping their abdomens, not their mouth parts, into the water. What are they doing?

A. You may have observed female dragonflies depositing their fertilized eggs into the water, where they will hatch into larvae. The eggs are sometimes inserted into underwater plant stems and sometimes smeared onto the surface of the water.

More than three hundred eggs may be laid at one time. Sometimes several hundred females lay their eggs in the same spot, producing large egg masses up to two feet across.

HIGH FLIES

Q. How high can a housefly fly? Are some insects better adapted than others to high flight?

A. How high an insect can fly depends on air temperature. Generally speaking, most insects can fly as long as they stay in air above 50 degrees Fahrenheit. If the air temperature at ground level is 90 degrees, the 50-degree border is around 6,000 feet; if ground temperatures are in the 70s, the ceiling is around 3,600 feet.

Insects have the ability to adjust their altitude, looking for the temperature they want to fly in. When it gets too cold, they tend to fold their wings and drop.

A number of insect species are adapted for long-range migration. Some of these species have large bodies and wings, but some that have very small ones, like leafhoppers, also migrate long distances. All seem to sense approaching storm fronts and ride the fast-moving winds ahead of them. They use energy to get up into the fast-moving air, then glide as the storm front moves them along.

The fast-moving air begins at the planetary boundary layer, about two and a half times the height of the largest surface obstruction around. So if there are 80-foot trees, it would begin at roughly 200 feet.

BITES IN THE SUNLIGHT

Q. Since I seldom see mosquitoes in daylight, can I avoid bites by going out only in the sunshine?
A. Unfortunately, no. There are many species of mosquitoes, perhaps 3,000 worldwide and 150 in the United States. The females, which seek a blood meal to produce eggs, have many different kinds of feeding habits.

While many of the mosquitoes that carry disease are

most active at dawn and dusk, even they may be roused to bite if their resting places are invaded in the bright daytime. Others may feed all day.

For example, the Asian tiger mosquito, *Aedes albopictus*, which is associated with the transmission of dengue, eastern equine encephalitis, and dog heartworm, is a very aggressive daytime biter, with peaks in early morning and late afternoon.

The dark ricefield mosquito, *Psorophora columbiae*, which in large numbers has been reported to kill livestock, has females that are furious biters, day or night. Luckily they prefer the blood of cattle to that of humans.

Aedes trivittatus, while not a major disease carrier, is a severe nuisance, because it attacks in swarms. Its bite is more irritating to humans than that of many other mosquitoes, and it does not hesitate to bite in bright sunlight when its haunts are invaded.

MORE OF A STING

Q. Is the toxicity of a scorpion's sting related to its size?

A. Scorpion experts have found that scorpion species with smaller and more slender claws typically have more toxic venom. But just as with bee stings, any scorpion's sting can be fatal to a person who is severely allergic to it.

The toxicity of the venom also varies by species and prey. A species' particular cocktail may contain many chemicals, some aimed at insects and some toxic only to mammals.

The different chemicals cause many different kinds of damage and symptoms. Some merely produce pain and swelling and are believed to be protective. Some, used for killing prey, cause severe neurological damage by blocking chemical channels in nerve cells. Some cause direct damage to the heart muscle. In humans, the toxic effects can lead to respiratory paralysis, extremely high body temperatures, dangerously low or high blood pressure, a rapid heartbeat, and other conditions.

The stings of most American scorpion species are not usually fatal; for example, the Arizona Poison and Drug Information Center says that only one of the thirty species in the state is regarded as life-threatening. But stings are much more perilous for children.

LONG LIVE THE SPIDER

Q. How long do spiders live?

A. With some exceptions, spiders in temperate regions have one-year life spans, with the females outliving the males by weeks or months.

Perhaps the most interesting exceptions belong to the family Lycosidae, or wolf spiders. Some members of this family live in elaborate silk-lined underground burrows. The females protect their eggs in sacs that they carry everywhere and care for assiduously. In burrowing species, the female may even carry the sac to the surface to be warmed by the sun.

After hatching, the spiderlings may spend the winter in the home burrow and may not reach full maturity until the second year, eventually moving out to build their own burrows.

The life span of a black widow spider depends on when it is hatched. Those hatched in late spring or early summer reach maturity in about four months, before cold weather, but may delay egg laying until spring and then die a few days later.

Those that emerge later may find their development severely retarded by the cold and the shortage of food. Only when warm weather arrives again do they mature. They may live a year and a half before reproducing. In the laboratory, some have lived through two or three summers.

THE CHIRPING THERMOMETER

Q. Do crickets chirp more slowly as the temperature drops in the fall?

A. In some species of cricket, the number of chirps per minute used by the male to attract females rises and falls along with the outside temperature, and can in fact be used as a rough thermometer.

Since insects are small (with a high surface-to-volume ratio) and cold-blooded, with relatively slow metabolisms unless they are exerting themselves, their temperatures reflect the outside temperature. Muscular activities

such as chirping tend to vary with that temperature.

The formulas for determining the surrounding temperature depend on counting the chirps in a certain number of seconds, usually 13 to 15, and adding a constant number, usually from 37 to

40. The most carefully studied species for this purpose is the widely distributed snowy tree cricket, *Oecanthus fultoni Walker*. Scientists have determined that by counting how often it chirps in 15 seconds and adding 40, a consistent approximation of the temperature in degrees Fahrenheit can be obtained.

Each species makes a different sound, and the chirping speed varies between species as well. There are silent crickets among the 3,000 or so known species (they use scent to attract mates) and even a few katydids that chirp faster to warm themselves up.

Source of the Flies

Q. When I bring fruit into my insect-free home, fruit flies seem to materialize. Are they drawn to the fruit or do they come in with it?

A. It is hard to be sure, but depending on when the flies show up, you can make an educated guess.

Flies have a particular amount of time to spend as egg, larva, and pupa before they climb out of the pupa. In the common fruit fly, this takes about eight to perhaps ten or twelve days. So if you buy a banana and bring it into your warm home and flies appear immediately or in two or three days, more than likely they came out of the banana skin. If the fruit sits around for ten days or so before flies appear, more than likely it drew flies that came in from outside.

The buyer would not see pupae embedded in the fruit's skin because they are only 2 to 4 millimeters (about 0.08 to 0.16 inches) long.

Q. What are dust mites? What can be done about them?

A. Dust mites are often referred to as tiny insects, but they are not. Instead, they are arachnids, part of the large group of eight-legged creatures that includes spiders and ticks. Several species of dust mite belong to the genus Dermatophagoides, a name that describes in Latin their way of life: They live mostly by devouring the skin cells that human beings shed.

Dust mites are very small; about 7,000 could fit on a dime. They set off allergic reactions when their feces and jettisoned external skeletons are inhaled. The proteins that cause the reaction are now being isolated in the hope of finding ways to fight allergies and asthma.

Dust mites thrive in human clothing, decorative fabrics, upholstery, and especially bedding. Warm, moist sleeping bodies provide them with an ideal environment.

Fighting mites involves cutting off their sources of warmth, food, and water, especially in the bedroom. Vacuum regularly, but wear a mask to avoid breathing in the dust you stir up; try an air filter. Remove rugs, curtains, upholstered chairs, and other fabric items. Keep mattresses and pillows in mite-proof plastic covers, and use synthetic pillows that can be washed rather than feather pillows. Launder clothes and bedclothes frequently, preferably in hot water.

Keep temperatures and humidity low, if possible, because dust mites prefer high humidity and temperatures above 68 degrees Fahrenheit.

CHAPTER 7

It's a Zoo Out There

How the Fittest Stay That Way

Q. How come zoo animals such as lions, not to mention my pet cat, seem to lie around all day and yet stay fit?

A. It is partly a matter of appearances, with zoo animals less active at peak visiting hours, and partly a matter of normal rhythms of carnivore life. Lions in the wild are normally inactive for twenty to twenty-two hours a day because they need to conserve their energy for hunting, never being quite sure where their next meal is coming from.

As for the house cat, although it does know when the next can will be opened, the basic behavior pattern is the same. Even a cat that has never seen real prey will stalk a butterfly through the window. Young animals can afford to do that, but from an evolutionary point of view, old lions especially must conserve their energy for the business of surviving. At some point in life, when the metabolism does begin to slow down, a sedentary cat may become overweight.

Of zoo animals, many are from hot climates and are most active in early morning and late afternoon, zoo-keepers say. They sensibly lie low from eleven to three on hot days.

Zoo animals have the benefit of a nutritional staff that prepares diets as close to natural as possible. Modern zookeepers also make sure that animals mimic natural behavior, living in groups, with plenty of space, and for-aging and competing for food. Some do tend to get chunky, like the dominant animal in a group that always gets its fill, but most stay lean.

GETTING THE MOST FROM GRASS

Q. People are told to eat a well-balanced diet, but if cows eat only grasses, how do they obtain all the nutri-ents their cells need?

A. All animals need various nutrients—things like amino acids and vitamins—to build and maintain their cells and to provide energy for bodily processes.

Carnivorous animals get many of these nutrients ready-made from the other animals they eat. Vegetarian animals get the nutrients not just by chemically changing plant material through enzy-matic processes but also from the action of beneficial fermentation microbes in their guts and from the bodies of the microbes themselves.

Omnivorous animals, like humans and pigs, can obtain their nutrition from both animal and plant food, or from either one.

In the cow, for one vegetarian example, there is a multi-stomach system. In the large, multi-compartmental forestomach, microorganisms ferment plant cellulose and release volatile fatty acids; the acids come into contact with cells with a large surface area that absorb them to provide much of the energy supply.

Later on in the digestive process, chewed and rechewed food enters a true stomach and intestines, more like those of humans, where it is mixed with more digestive juices that change the chemistry of the food. There, amino acids and other nutrients are absorbed and the bodies of the microorganisms from the forestomach are also digested.

THE MIGHTY WOLVERINE

Q. A nature TV program showed a wolverine driving three much larger wolves away from its kill. How is this possible?

A. Pound for pound, a 40-pound wolverine, formally *Gulo gulo*, is simply stronger than many of its adversaries and is capable of killing animals much larger than itself. Wolves are probably aware of this.

For example, a wolverine's jaws are so strong that it can completely crush a frozen caribou's large bones, part of its adaptation to life in a wintry climate. It can also use its relatively large flat feet like snowshoes in the efficient pursuit of hooved animals over soft, deep snow. Its main food in winter is reindeer or caribou, which it may kill or scavenge. It also devours smaller rodents and occasionally berries.

The wolverine is the largest member of the weasel, or Mustelidae, family, weighing up to 55 pounds. In Alaska, it averages about 33 pounds for males and 22 for females. It is widely distributed in the arctic and subarctic tundra.

Because of its ferocious and wily nature, it has been comparatively poorly studied, and the status of the wolverine throughout its range is largely unknown, according to the Wolverine Foundation, with headquarters in Kuna, Idaho. The foundation was started in 1996 to promote interest in the wolverine's status and ecological role.

To Sleep, Perchance to Hibernate

Q. Help settle a family dispute: Do bears hibernate or not?

A. Bears do hibernate, though not as deeply as some other hibernating animals.

Deep hibernators spend most of the winter with drastically lowered body temperatures. Bears lower their body temperatures only slightly and continue to burn about 4,000 calories a day, resulting in a Zen-like state of watchful rest. They can quickly rouse themselves in response to intruders and can even nurse their young during hibernation.

But bears' state of hibernation, which is also called denning, can outlast that of deeper hibernators, like the ground squirrel. Ground squirrels do not sleep straight through the winter but rouse themselves every couple of weeks to urinate or defecate, presumably with complex metabolic triggers. Bears can spend up to five months without eating, drinking, defecating, or urinating.

Scientists now suggest that recycling is what permits bears to engage in what seems like a physiological impossibility and emerge healthy. They seem to be able to use products of bone degradation to build new bone and to use urinary wastes to make protein.

BEAR PAIRS

Q. Can the polar bear mate with the brown bear to produce fertile offspring? Is the polar bear *Ursus maritimus* or *Thalarctos maritimus?*

A. The polar bear can breed with the brown bear, and the offspring would probably be fertile. But it would never happen in the wild. The ranges of polar and brown bears do not overlap, and even if two met, they would see each other as competitors, not mates.

Such interbreeding has occurred in captivity, but not in more than two decades. In the early 1980s, an Asiatic black bear and a spectacled bear interbred, but the male offspring was sterile. Modern-day zoo management frowns on these things.

The polar bear has been formally called both *Ursus maritimus* and *Thalarctos maritimus. Ursus maritimus,* meaning sea bear, was originated in 1774 by Commander C. J. Phipps, a British naval officer and author of *A Voyage Towards the North Pole*. The more recent designation *Thalarctos maritimus* combines the Greek *thalasso*, or sea, and *arctos*, or bear of the north. *Arctic* itself means "Country of the Great Bear," but the name comes from a bear in the night sky, the Great Bear constellation, or Arktos.

ASLEEP ON THEIR FEET

Q. Why do horses sleep standing up?

A. Horses can doze standing up and spend more time upright than other animals, but get their deepest rest, the so-called REM sleep, only when lying down. In total relaxation, which rapid eye movement or dream sleep involves, the system of tendons and ligaments that keeps horses' legs extended does not work. If a horse goes into REM sleep while it is standing, it tends to fall on its knees.

Wild horses lie down less, a fact that some researchers attribute to the need to be ready to flee predators. But others suggest they lie down less because during certain times of the year, notably

winter, they eat day and night because less feed is available.

In summer, wild horses lie down a fair amount. When they do lie down, one usually remains standing as a sort of guard.

Reindeers' Close Relations

Q. Are caribou the same as reindeer?

A. They are very closely related, so closely that they are now usually considered different races of the same species, *Rangifer tarandus*. Even the somewhat smaller domesticated reindeer are part of the same group.

Caribou coloration varies widely, from nearly black to brown to gray to almost white, and some populations migrate hundreds of miles, with some preferring the arctic tundra for summer and the edges of woodlands for winter. Their coats may vary from mostly brown in summer to gray in winter.

Caribou or reindeer range from Scandinavia, European Russia, and Siberia to Greenland, Alaska, and Canada. They have even been introduced into the island of South Georgia in the South Atlantic.

There are nine races or subspecies of caribou or reindeer; many of the European races are domesticated.

Reindeer are the only deer in which both sexes have antlers, which they shed at different times of the year. It has been suggested that the females' antlers help them compete for food in the unusual herds that reindeers form, which are both large, with up to forty females per male, and coed, though adult males are often solitary until it comes time to gather a harem.

Virtuoso Builders

Q. Why do beavers build dams?

A. Beavers make their complex domestic arrangements to improve security from predators and to move and store a reliable food supply. Dam building is only one of their construction activities, along with canal building and lodge building.

The dammed-up ponds make an important contribution to the beavers' well-being by creating storage at the right depth to stash a winter's supply of bark, twigs, roots, and leaves at temperatures above freezing, but cool enough to keep their nutrient value intact. The ponds also improve security and must be deep enough for a family of beavers to swim under winter ice from their lodge to the hidden food.

The dams are constructed by moving mud, stones, sticks, and branches across a shallow stream. Beavers use their forepaws to move mud and small stones from the stream bottom to the dam site. Large sticks and branches are towed to the site with the beavers' large incisors and are used to support the dam and keep the other components in place.

The females are the most active in dam building, though all adults and yearlings may take part. Most construction takes place at night.

CHATTY CHIPMUNKS

Q. Why do chipmunks sometimes chirp loudly and continuously for fifteen to twenty minutes at a time? It doesn't seem to have anything to do with food, as the ones in our backyard do this even when well fed.

A. As you suspect, the noise probably has less to do with foraging, as the familiar fat-cheeked eastern chipmunk (*Tamias striatus*) can easily collect and store a lifetime supply of food in a single season, than with protecting territory and warning off interlopers.

Chipmunks have a variety of vocal sounds, not all of them translated by scientists, but the one you describe, which has also been rendered "chip-chip-chip" or "chuck-chuck-chuck" or "chipp-R-R-R," is often associated with driving away an intruder perceived as threatening the burrow. The high-pitched sound is repeated every few seconds and may be echoed by a chorus of concerned neighbors.

The scolding is probably being delivered by females, as they are apparently even more territorial than the males. But neither sex is particularly gregarious, except for the period around mating, when the couple may play and vocalize together for some time. Shortly afterward, the male is vociferously driven off. The female can have two litters a year, and the young are, in turn, loudly urged to leave home in six to eight weeks.

CALLING A SNAKE A SNAKE

Q. Do the terms *snake*, *viper*, and *serpent* have scientific differences?

A. The subject is a confusing one because of the rather inexact use of terms in most post-Roman cultures. Especially large or poisonous snakes are commonly called serpents, and biblical tradition calls the shape

Satan assumed to tempt Eve a serpent. But technically speaking, all snakes can be called serpents, from the Latin *serpere*, meaning to creep.

However, scientists use the word *snake* to mean any member of the taxon Serpentes, generally considered a suborder of the order Squamata, although Serpentes actually sits closest to monitor lizards in the evolutionary tree of lizards. (Snakes, then, are really just one legless subset of lizards.)

While the word *serpent* may be used as a synonym for any snake, a viper is any member of the solenoglyph snakes, including rattlesnakes, adders, water moccasins, and copperheads.

Generally, systematic herpetologists place vipers in two families. They are the Viperidae, or Old World vipers (adders, puff adders, etc.), which lack heat sensory pits on the face, and the Crotalidae (pit vipers), found in southeastern Asia and the Americas, which do have sensory pits on the face, including the rattlesnakes, bushmasters, fer-de-lance, cottonmouths, and tree vipers.

A Meal in One Gulp

Q. How can a snake swallow prey that is bigger in diameter than it is, such as a whole rabbit?

A. Most snakes have specialized body structures that let them swallow things that are larger than their heads or necks. First, a snake's skin is very elastic. Even more important is the fact that the snake's jaws are made up of four elements, not two, and they are only loosely attached with ligaments that can stretch during feeding.

The two sides of the upper jaw are independent, and

at the back of the bottom jaw, where there is a single hingelike joint in other animals, a snake has an extra bone that drops down so the two parts of the jaw can open to allow prey to pass into the neck.

A snake's ribs are also flexible, not fused to a sternum as they are in human beings, so they, too, can spread and allow food to pass down into the stomach. Finally, the whole process is pushed along by a complex group of strong muscles.

Some large snakes that are 12 to 15 feet long have no trouble handling a 15-pound rabbit. About 98 percent of snakes can do similar tricks with prey appropriate to their size, but a few underground snakes that feed on insects and soft-bodied invertebrates have lost the ability.

UP A CREEK WITH NO LEGS

Q. Can rattlesnakes swim? I need to know before I take a dip in a lake bordering on Rattlesnake Hill.

A. Rattlesnakes are good swimmers, though they are not in a class with species that spend all or most of their lives in water or even with the familiar garter snake. From anecdotal accounts, we know that rattlesnakes will swim readily enough when their travels in search of food, mates, or refuge require them to cross water.

Rattlesnakes are buoyant and float well, and they propel themselves fairly fast in water with a sinuous motion. They may hold their rattles above water when resting or floating, but they use their tails for propulsion, with their rattles below the surface.

Although rattlesnakes can bite when in or under water, they cannot strike efficiently because they lack a solid base. You'd have to be pretty careless to get close enough to a swimming rattler to be within striking range and then get it mad enough to bite.

CHAMELEON COLORS

Q. How and why does a chameleon change its colors?
A. There are two groups of lizards commonly called chameleons, and they change colors in different ways.

True chameleons belong to an Old World family called Chamaeleontidae, with two genuses and about eight-five species. They have a variety of colorations, which they can change, to a greater or lesser extent, by concentrating or spreading out the pigmentation in skin cells, called melanophores.

The process is under the control of the autonomic nervous system, and the change is not just a matter of color matching. Rather, it is a complex response to a variety of conditions, including light and temperature, but also "emotional" factors such as fear or triumph in battle with another chameleon.

The strength of the light is apparently the most important factor, and for certain species, this sometimes results in camouflage. For example, the flap-necked chameleon, *Chamaeleo dilepis*, an African species, turns green when among leaves and yellow or reddish-brown when on bark. However, in response to a threat, it turns blackish-green, with yellow and white spots.

Members of the other group sometimes called chameleons, anoles (which are New World relatives of the iguana in the genus Anolis), have a limited color-change system under the control of hormones rather than nerves.

THE FERTILE TURTLE

Q. How do turtles reproduce?

A.   *The turtle lives 'twixt plated decks*
            Which practically conceal its sex.
            I think it clever of the turtle
            In such a fix to be so fertile.

The poet Ogden Nash was right about the turtle's external ambiguity and fertility. Turtles are not only enthusiastic breeders, they also have external sexual characteristics that often make it hard for creatures other than turtles to determine which is which.

The male is sometimes distinguishable by an indentation or curvature in its plastron, or lower shell, which fits over the back of the female; females have a flat or convex plastron. To fertilize the female's eggs, the male turtle conceals a sexual organ inside the cloaca, or waste removal chamber. The male positions itself over the female and often grasps the upper shell, or carapace, with its claws, then curves its tail until the vent contacts the female's vent; the penis emerges for fertilization. The often dozens of eggs develop internally and are then usually laid and buried in sandy soil.

Fertilization is sometimes preceded by elaborate courtship rituals, with hours of demonstration followed by a few minutes of copulation. The female can store sperm to fertilize its eggs, sometimes years later.

FROGS' FLYPAPER TONGUES

Q. What is the sticky stuff on a frog's tongue, and how does it work to let frogs catch prey?

A. There is no special name for the sticky substance, but scientists do know it is secreted by mucous glands at the moment the frog's tongue hits the prey. Otherwise it would stick the frog's mouth together.

The tongue of a frog, which is hinged at the front of the mouth, not the back, can be envisioned as working like a human arm with the fist folded up to the shoulder.

When the "arm" is flipped out straight, the part of the tongue that hits the prey corresponds to the back of the hand, not the palm. Where it touches the prey, muscles in the tip contract so that it instantly exudes a sticky substance. Then, like an arm folding back up to the shoulder, it carries the prey back into the mouth.

The tongue moves in and out as fast as the blink of an eye. Then the eyes are depressed by a further set of muscles, which forces the food backward into the mouth. Once it is in the throat, hairlike projections called cilia move it toward the gut.

Some aquatic frogs have evolved with no tongues. They never come out of the water and move food into their mouths using their limbs.

FLIGHT OF THE FISH

Q. Why do flying fish fly?

A. The several species of oceanic flying fish seem to take flight either when they are alarmed or when they are attracted by a light shining in the night.

Most flying fishes belong to the family Exocoetidae and live in the surface waters of all tropical oceans. They are able to leap out of the water and glide above it because they have overdeveloped pectoral fins and in some cases pelvic fins as well.

The flying fish that have two pairs of "wings" build up swimming speed much as an airplane taxis before take-off; they gather speed at the surface by sweeping the tail fin rapidly from side to side, vibrating it like an outboard motor. Having reached speeds that sometimes exceed 35 miles an hour, they finally launch themselves into the air. Flights last from about ten seconds to as long as thirty seconds and may carry the fish 650 feet or more.

Flying fish species range from about 5½ inches to 20 inches in length. They are often good eating, and one technique of fishing for them relies on their attraction to light. A fisherman may simply anchor a canoe with a foot or two of water in it and a lantern hanging over it. The next morning, the boat may be full of fresh fish trapped in water too shallow for them to make a take-off.

Q. Is it true that some fish are both male and female?

A. Yes, some fish are simultaneously hermaphroditic, possessing the reproductive equipment of both sexes at one time; a few can fertilize their own eggs. In some species, called sequential hermaphrodites, an individual can change from male to female, from female to male, or back and forth. Some fish species are entirely female and reproduce by cloning their own DNA, though the process is triggered by sperm from males of closely related species.

Conversion from male to female is called protandry; conversion from female to male is called protogyny.

Sex changes in fish were once thought to be extremely rare. It is now known that changes from female to male occur in at least fourteen families and changes from male to female in at least eight.

In a study of a reef fish called the bluehead wrasse, reported in 1984, sexual identity and behavior were found to vary with the density of the population and the size of the reef. For example, sometimes a large dominant male fertilized the eggs of many females, and sometimes fish took turns releasing eggs to be fertilized by sperm of other fish, which in turn released eggs for fertilization.

In another wrasse species, a solitary male fertilized the eggs of several females; if the male died, one of the harem became male.

In some hermaphroditic species, the adaptation is thought to help ensure reproductive success when other members of the species are few and far between.

Changes in the sex of the fish are presumed to be influenced by hormones, but the mechanism is not fully understood.

TELLING THE DOLPHINS FROM THE PORPOISES

Q. What is the difference between a porpoise and a dolphin?

A. They are cousins but belong to different families. All are members of the cetacean order, which includes whales, and all belong to the suborder odontoceti, literally meaning toothed whales.

The thirty-two species of seagoing dolphins (members of the family Delphinadae) are perhaps most easily distinguished from the six species of porpoises (members of the family Phocoenidae) by the fact that most dolphins have a well-developed beak and a bulbous forehead. Dolphin bodies also tend to be more slender and streamlined than the whalelike porpoises. River dolphins in the family Platanistidae, which inhabit muddy rivers, have even longer, more slender beaks, prominent forehead pads, and small, weak eyes.

The bulging forehead of the dolphin houses a pad of fat called the melon, which is thought to help focus the high-frequency sonar beams with which dolphins navigate and find their food.

Orca, the killer whale of motion picture fame, *Orcinus orca*, is actually a kind of dolphin, but one that lacks a pronounced snout. Performing dolphins are Atlantic bottle-nosed dolphins, *Tursiops truncatus*.

Q. Is a third whale involved in positioning whales for mating?

A. In some whale species, a third whale may have a role in courtship, but it is now thought to involve competition, not assistance.

The California gray whale, *Eschrichtius robustus*, has many more males than females. During courting, two

males will often follow one female, and the trio will stay together right up to the moment of copulation. The female slows down, signaling a willingness to mate, and the males use their flippers to slow her down even more. The males swim upside down beside her, each trying to mate. When one succeeds in rotating his underside alongside that of the female, copulation occurs.

There was some speculation that an extra male, or males, held the mating pair together to brace them for mating. However, this idea is now mostly dismissed as a misinterpretation of what is going on in a process that is difficult to study.

As for getting into position to mate, the female usually floats right side up and the male comes up from below. What you do see in mating whales is a lot of rolling around.

Sex and the Single Fossil

Q. How can scientists tell what sex a fossil is, or was, when they have only one example?

A. Although it can be difficult to know with certainty, there are clues in the shape and size of bony landmarks on the skull. You usually can't sex other bones reliably, except for the pelvis.

With the skull, what scientists look for is the difference between what is called robust morphology, meaning typically heavier, larger shapes and sizes, and gracile morphology, meaning smaller and more "graceful" shapes. Males tend to be more robust in the browridge, the mastoid process behind the ear and the lower jaw.

But with only one fossil example, researchers must rely on what they know of the species and make an educated guess. Often it can be unclear whether researchers are looking at a species that has a marked difference in size between the sexes, called sexual dimorphism, or an entirely different species.

The Species Census

Q. How many living species are there? What is the breakdown of plants, vertebrates, etc.?

A. One recent estimate is 10 million to 13 million living species. A 1988 study said only about 1.5 million species then had scientific names, though far more certainly exist. The study pointed out that "total" figures are not actual counts, but theoretical reckonings based on things like the number of species that make up typical food chains in different climates. Some groups, like mammals, have been intensively studied, and others, like mites, are little known. A big problem is defining a species: some biologists discern two hundred kinds of British blackberry, for example; others might list twenty, or two or three.

Two kingdoms, the prokaryotic monerans and the eukaryotic protists (two kinds of microscopic unicellular organisms), make up about 5 percent of recorded living species; the kingdoms of funguses and plants make up about 22 percent; the rest are animals.

In one estimate of the animal kingdom, well over half the total species are insects; mammals make up 0.388 percent of the total, and other chordates (animals with at last a precursor of a spinal cord) are just over 3.7 percent.

CHAPTER 8

Secrets of Cats and Dogs

ITCHY ANIMALS

Q. You hear all the time about people who are allergic to cats and dogs. Do cats and dogs ever develop allergies?

A. Yes, allergies are common in cats and dogs. They arise by similar mechanisms and can cause similar symptoms, though skin eruptions are more common in cats and dogs than sneezing or digestive distress. Treatments may be similar, too.

In an allergy, the animal's immune system is hypersensitive to some substances it encounters, called allergens. For example, it is very common for pets to be hypersensitive to certain proteins in the saliva of fleas.

Pets, like people, can be allergic to substances that are airborne and inhaled, like pollen, mold, and dust; to substances that they come into contact with, like wool or flea collars; and to foods, like corn, wheat, soy, beef, or dairy products. The allergic tendency can be inherited or develop after repeated exposure.

It would be very unlikely for a cat to become allergic to its owner, since humans do not ordinarily lick themselves clean the way cats do, spreading some allergy-causing protein in their saliva onto fur and dander that is then blown about in the air.

Treating a pet's allergy may involve protecting the animal from re-exposure to the allergen, desensitization therapy, shampoos and rinses to remove an irritant from the skin, and the use of antihistamines and steroids.

A CAT'S MAD DASH

Q. Why do cats periodically run around madly as if they were chasing a fire engine?

A. It is probably more evolutionary than it is anything physiological, and is possibly related to hunting techniques. It is not usually seen in farm cats, who have to work for a living. Their predatory behavior involves sitting beside a mouse hole for hours, then pouncing.

Instead, the mad dash is often seen among well-fed,

secure, happy house cats. Cats are crepuscular animals that hunt at dawn and dusk, and the running around often coincides with the time they would ordinarily be hunting.

It may also be a form of play. Cats often seem to be stimulating themselves, similar to a child on a swing. If all the dashing around is a problem, veterinarians suggest trying interactive play with a pouncing toy. That might reduce the fire-engine behavior.

FELINE ANTENNAE

Q. What are cats' whiskers for?

A. Whiskers give cats important sensory information and also have a role in communication between cats. They are far more important to a cat than a dog's whiskers are to a dog, and they should never be trimmed, as they tell a cat whether it can squeeze through a narrow passage, for example.

Technically called vibrissae, whiskers or tactile hairs are clustered in three major groups on the sides of the cheeks, above the eyes, and farther back on the cheeks. There are also a few on the chin and shorter ones on the back of the front legs.

Stiffer and at least twice as thick as ordinary hairs, vibrissae have roots that go three times as deep in the skin and are surrounded by nerves and blood vessels. The slightest motion of a hair, even from a breath of air, triggers a sensation.

Vibrissae are helpful to a cat maneuvering in dark areas, especially when hunting, and when a cat has prey in its cupped paws, ready to move it toward the mouth, the hairs give signals of the next meal's exact position.

As for communication between cats, whiskers pushed forward may indicate friendliness or inquisitiveness, but when they are pulled back flat against the cheeks, they can signal either aggressive or defensive hostility.

SHOCKING BEHAVIOR

Q. How can I avoid shocking my cat with static electricity when I pet her in cold, dry weather?

A. There are two approaches. One is avoiding the buildup of the charge; the other is discharging it so you cannot shock the cat or yourself. (The cat also builds up a charge, but because of its smaller size, most of the flow is from you to the cat.)

In friction between any two materials with different affinities for electrons, one winds up negatively charged, with a surplus of electrons, and one positively charged, with a deficit. The more different the affinities, the greater the charge. Either a surplus or a deficit can make a spark.

One simple way to discharge a human buildup from walking across a rug with rubber-soled shoes, for example, is to touch a water pipe, faucet, or other grounding device just before touching the cat. To prevent a strong shock to yourself, touch the keys from your pocket to the pipe. You will still feel something, but the sharpest part of the spark is at the point of the key. The discharge tends to occur at the minimum radius of curvature.

There are also antistatic sprays, some for preventing static cling in garments by insulating the fibers and some for use around computers that spread the charge around and allow the discharge of electrons to leak out over a wide area.

In either case, be sure to spray your pant legs and carpet, not the cat.

A KIDNEY FOR THE CAT

Q. Can cats have organ transplants?

A. Yes, in some cases. Kidney transplants were pioneered by the School of Veterinary Medicine at the University of California at Davis and are now performed at some other veterinary colleges around the country. They can cost $3,000 or more.

There are no other transplant programs for dogs or cats, though there was initial research on partial liver transplants.

Because there is no organ procurement program for animals, veterinarians have to use live donors that can survive donation of one of two kidneys. At Davis and other schools that have transplant programs, the school maintains a colony of donor cats, and the owner of a cat that receives a kidney has to agree to adopt the donor.

No need has been found for tissue matching in cats, and rejection of a transplanted kidney can easily be suppressed with cyclosporine. Without the drug, rejection occurs in about twenty-one days. The school at Davis also has a dialysis unit. Cats with acute kidney failure, caused by things like poisoning, usually need three to four weeks of dialysis, cost-

ing $3,000 to $6,000. In cats with chronic kidney failure, a few weeks of dialysis before a transplant makes success more likely.

Sensing a Storm

Q. Why can my dog sense when a storm is coming an hour before it arrives?

A. "Sense" in its physical meaning is the key word. Many case reports cite instances of animals' behaving in strange ways before natural phenomena, but there is little scientific data.

Domestic animals, dogs and cats in particular, have hearing and other sensory capacities that are greater than those of people. Their hearing is more sensitive and covers a wider frequency range, so it wouldn't be surprising if they could hear the sounds of an approaching storm.

Dogs and cats also have a much better sense of smell. Meteorologists say that storm winds may carry smells from other areas and temperature and humidity differences that are detectable far in advance of any actual precipitation. Even people sometimes say they can smell when snow is coming. That is especially true of certain kinds of heavy snowstorms accompanied by winds from the south.

Fido's Brain

Q. If a Chihuahua and a mastiff have such different head sizes, does their brain size vary that widely, too, and does it affect their comparative intelligence?

A. Brain size is not generally seen as a reflection of intelligence in dogs or in other species. That is evident if you compare other species with different brain sizes—for example, elephants and people: People are far more intelligent despite the huge disparity in physical size.

Intelligence has more to do with the development of different areas of the brain over the course of evolution and the devotion of different areas to specific functions. The potential development of the brain is equal among the various breeds.

However, not all animals develop equally regarding genetics and environmental influences. There could be marked differences in intelligence between dogs of nearly identical breeding, even littermates. And as any Chihuahua owner could tell you, some of the smallest dogs can be some of the craftiest.

Dogs were selectively bred over millennia for different kinds of intelligence and behavior. People who have tested dogs with obstacle courses and other measures of talent have found that some breeds excel in particular areas, such as herding or retrieving. Some, like the border collie, German shepherd, and poodle, seem to excel in many fields.

GARRULOUS PARROTS

Q. Do parrots and other birds that talk have any special physical equipment in the larynx or tongue that helps them speak?

A. There is no special equipment, just a highly developed ability to imitate. The ability to imitate a species' own song is present in all songbirds.

Some birds, like wood-peckers or chickens, do not actually learn songs. But of the birds that do, some learn a little and some learn a lot.

Among the star imitators are parrots, mockingbirds, mynahs, and starlings. If these birds spend a lot of time with a person, they will learn a lot from the person, although it is not certain what brain structures are involved in the ability to imitate human speech.

The brain structures by which birds learn songs, however, have been largely mapped. Other research found that the series of structures informally called the song system is larger and more complex in birds that learn more complex songs than it is in those that can learn only simpler songs.

GUINEA PIG IDENTITY CRISIS

Q. Are guinea pigs rodents?
A. Probably, based on structural features and fossil evidence, though some scientists studying DNA evidence suspect otherwise.

There are studies, primarily of mitochondrial genes, suggesting that guinea pigs might not be rodents, but might belong to their own separate order. (Mitochondrial DNA is carried by the structures in cells that act as cellular power plants. It is inherited in a manner different from ordinary DNA.)

However, most of these studies were based on a limited sample of genes in a limited number of orders. More gene sequences should be studied for more mammals before we can be sure of the answer.

Another recent analysis showed that by using the same data as some of these studies, but employing another commonly accepted model of gene evolution, guinea pigs can be placed in the rodent group.

Finally, the traditional kind of evidence in mammal evolution is very strong that guinea pigs belong in the order Rodentia, both from morphology (structural features such as the head and the development of the gnawing teeth) and from paleontology (fossils from Central Asia).

CHAPTER 9

Animal Aviators

BEYOND LUNGS

Q. How do birds breathe? Is it more like mammals or like fish?

A. Like mammals, birds take in air and extract oxygen from it to supply body tissues, while fish respiration depends on getting the oxygen dissolved in water through organs called gills.

But bird respiration is significantly different from that

of mammals. In the very efficient avian system, the lungs are small, taking up just 2 percent of body volume.

The external respiration system uses the lungs and a much larger series of air sacs distributed throughout the body so that the oxygen supply communicates directly with body parts. Oxygen is taken in and exchanged for carbon dioxide waste in the blood, then the carbon dioxide is moved out; so far, much like mammals.

But in birds, the air flow is one way, through tubes and chambers, rather than two ways, in and out of the lungs, as it is in mammals. And when a bird inhales, air actually leaves its lungs; fresh air enters the lungs as it exhales.

First, a constant stream of air enters through the back of the mouth and goes into the larynx and trachea. It is channeled in a pair of tubes straight through the lungs and into the system of air sacs and hollow bones. Gases from the air sacs are sent forward through the lungs again, then through more air sacs into the trachea, and then finally expelled from the body.

SENSING SOMETHING IN THE AIR

Q. Do birds have a sense of smell?
A. Yes, and though it is highly variable from species to species, research is leading to a higher estimation of birds' smelling abilities.

Sensitivity to odors varies among orders of birds with the size of the olfactory bulb in the brain relative to that of the cerebrum. The bulb tends to be small, but it is well developed in some birds, especially those that fly and hunt by night, and research has found that some birds with relatively small olfactory bulbs can smell well enough to detect certain odors.

Scientists now suspect that most birds can use the sense of smell in daily activities, and individual species have been found to have high sensitivities adapted to needs like mating (finding a female in season by detecting gland secretions), feeding (smelling carrion or worms), or even finding their nesting burrows.

Some birds can smell just a trace of a substance that might lead them to food. For example, bacon fat poured on the surface of the ocean has been known to attract black-footed albatrosses from more than 18 miles away. Turkey vultures have been used to find leaks in a 42-mile-long oil pipeline; when ethyl mercaptan, which smells like rotting meat, was pumped through, the birds gathered at the leaks.

INVASION OF THE CROWS

Q. How can we get rid of the crows that have invaded the big garden in our city apartment complex, where we used to get songbirds?

A. Short of exterminating all the crows in town, there is not much chance of getting rid of them. Driving out one family simply opens a place for another young male to move in.

One possible way is to make things less attractive, but what attracts crows is the same thing that makes gardens attractive to humans. Crows like open green space, and a big lawn with a few trees is ideal. The lawn provides food like earthworms, mole crickets, and June beetle larvae.

If you don't want to cut the trees down and pave the lawn, another possibility is scaring the crows, who tend to be a bit more wary than other birds. A few disposable aluminum pie plates on a string might just change things enough so that the crows don't want to come into a particular yard.

Crows have moved into urban areas around the country in the last thirty years, becoming urban rather than rural birds. Family groups average three birds but range up to fifteen or twenty, many of them young males waiting for the opportunity to breed. Males usually breed first at the age of 5, females at 3½ to 4.

As for the songbirds that disappeared, the crows probably did not play a direct role, and the cardinals and others are probably still around.

HIGH-SPEED DIVE

Q. What bird flies the fastest?

A. The winner, wings down, is the peregrine falcon, *Falco peregrinus,* which may also be the fastest animal of any kind on earth. Estimates of its top speed when diving start around 200 miles per hour, and some authorities put it considerably higher, perhaps 275 m.p.h. Its top horizontal speed is 60 m.p.h. or more.

The high speeds are essential to the normal feeding habits of the peregrine falcon, which overtakes and kills other birds or bats in midair.
The bird cruises for hours, then plunges in a move known as a stoop. The prey is often killed by violent impact with the peregrine's snatching talons.

The peregrine falcon is about 15 to 21 inches long,

and its wingspan is about 36 to 44 inches. The population of peregrine falcons has made a well-publicized recovery since the pesticide DDT was banned in 1972. The residue of the poison in the peregrines' prey built up in the fatty tissue in the parents, and the female falcons laid eggs that were deficient in calcium and too weak to withstand the incubation of thirty-two days or longer under a mother who weighs about three pounds.

Staying in the Pink

Q. Is it true that flamingos are pink only because they eat pink shellfish?

A. The pinkness of flamingos is determined by food, but not by pink shellfish. The factors in the flamingo diet that ensure pinkness are carotenoid pigments, which are found in plankton, diatoms, and blue-green algae that the birds strain out of the muck in which they feed. The birds process yellow carotene into a red compound, canthaxanthin, which is stored in their legs and feathers.

If flamingos do not get enough of the right pigment, they lose their color when they molt. The color is important, because flamingos do not seem to breed successfully without it. In captivity, they were once fed ground-up carrots and red pepper to keep them pink, but now zookeepers try to reproduce their natural diet or give them synthetic canthaxanthin.

Flamingos are not the only birds that change color with diet. "Red" canaries may be fed special nutrients to keep them red, though bird shows may frown on this color feeding. Cardinals also fade noticeably in winter, when they do not have access to their normal wide variety of foods.

Flamingos have very unusual feeding equipment. They eat with their heads upside down. The lower jaw contains a thick, fleshy tongue that moves back and forth to create powerful suction. Both upper and lower jaws contain rows of projections called lamellae that meet to create a strainer to filter out edibles as water is pumped past them by the tongue. The projections range in size from large hooks to a Velcro-like fringe.

The greater flamingo, with larger lamellae, strains out and swallows small invertebrates, while the lesser flamingo, with a finer sieve, captures blue-green algae. The two species can feed side by side and eat different foods.

SOARING INTO OLD AGE

Q. If I had a pet bald eagle, which I could never have, of course, how long could I expect it to live?

A. Smaller eagles have occasionally been "domesticated" by falconers and were traditionally reserved for the hunting pleasure of medieval kings.

The American bald eagle, which has a wingspan of up to 7½ feet and would therefore require a very large falconer, has been reliably reported to have lived for 30 years and 5 months in captivity at the National Zoo in Washington. One banded bird in the wild in Mexico lived 10 years and 5 months and might have lived much longer but was shot.

A bald eagle, formally *Haliaeetus leucocephalus*, does not even develop its distinctive white-feathered helmet until its fourth year. Before that, it goes through a series of immature plumages, which are wholly or partly replaced by a series of molts, or feather losses.

While the bald eagle's head is not bald, but fully feathered, the legs are unfeathered, and the bare-legged look helps bird-watchers distinguish between immature bald eagles and immature golden eagles.

TURNING ON THE FLY

Q. How do birds turn in flight?

A. The assumption that a bird's tail functions only as a horizontal surface is erroneous. A bird also can twist its tail from left to right and downward, so it acts almost as a braking system that allows it to bank into turns.

Some birds are better at this than others. Some hawks in the group known as accipiters, which spend most of the time in forested settings, use their tails as rudders so

they can rapidly weave their way through trees without striking them.

There are other turning mechanisms as well. The large webbed feet of waterfowl are used to steer during landing by creating wind resistance.

Birds also steer by changing their wing configuration. They can change the angle at which the wing intersects the stream of air, known as the angle of attack. A greater angle of attack creates more wind resistance, and applying this differentially on each wing allows the bird to turn.

The Birds and the Bees and Birds

Q. How do birds have sexual intercourse?

A. Parents have supposedly explained human reproduction to children in terms of the birds and the bees; if birds were really used to make this explanation, it would be widely believed that fertilization occurs after the briefest period of proximity of the waste elimination organs.

Most birds do not have external sex organs. The transfer of sperm from male to female usually takes place while the male bird is standing or treading on the female's back. The male twists his tail under that of the female so that the opening of the male bird's cloaca, or waste storage chamber, is close to that of the female's.

Sperm transfer takes place when both cloacae open outward. Tiny projections called papillae protrude from the back walls of the male's sperm sacs into his cloaca. These are brought into contact with the opening of the female's oviduct, or egg tube (most female birds have only one that works), and the sperm is released to fertilize the eggs.

Ornithologists usually describe the contact as a "cloacal kiss." It commonly lasts one or two seconds.

Bird Bombs Away

Q. Why are bird droppings whitish or spotted with white?

A. Bird droppings combine the whitish waste products processed by the liver and kidneys with the darker wastes that come from the digestive tract. In birds, the nitrogen-rich wastes are turned into a white-colored paste that is composed mostly of urates. The urates often form a separate white blob, as anyone who has worn a new hat outdoors at the wrong time has probably noticed.

The excretory system of birds is different from that of mammals because it takes up a minimum of space and requires a minimum of water for processing.

In mammals, nitrogen waste is turned into urea, a toxic substance that requires large amounts of water to flush it safely from the system and a urinary bladder as a holding tank for all the liquid. The urates excreted by birds, on the other hand, are concentrated but not toxic, and most birds do not have bladders to hold liquid to flush it out.

WHERE THE GULLS ARE

Q. I have observed gulls perched on all sorts of structures, but never on a tree. Why?

A. Most gulls do not have feet that are adapted to perching in trees and clinging to tree branches. Most have slightly webbed feet that help them move about in the water, so when they are on land, they do better walking on rocks or soft surfaces like sand.

Pure probability also puts gulls on the structures of civilization. Birds like gulls and crows have done quite well adapting to human-created environments, and the nice flat surfaces associated with human habitation—pilings, piers, and lighthouses, to name just a few—are ideal places to perch.

Gulls that can nest in trees, like mew gulls, have feet that are not fully webbed. Mew gulls make their nests fairly low, in spruce trees, not deciduous trees, where there is more surface area to cling to.

Many forest birds, in contrast with gulls, have feet with tendons that keep their toes automatically wrapped around branches. For these birds, it takes an effort to unclench.

HARD-HEADS

Q. Why don't woodpeckers damage their brains?

A. Woodpeckers like the acorn woodpecker of California, which hits the wood with its bill at speeds of 12 to 15 miles an hour, manage to keep their wits because their brains are held firmly in a case, which acts as a shock absorber. The bird's body also moves in a single plane like a metronome, avoiding the rotational forces that would twist and tear its brain loose from its moorings.

The human brain does not enjoy this kind of protection. Instead, it is encased in a sac of fluid. Thus when a cyclist hits the pavement, for example, the impact can make the brain bounce around in its sac, hitting first one side of the skull and then the other. That motion can tear nerve fibers and cause permanent damage.

CHAPTER 10

How Humans Get That Way

THERE MUST BE A REASON

Q. Is there an evolutionary reason for morning sickness?

A. Evolutionary biologists have suggested that morning sickness is a mechanism to help women avoid noxious substances that might harm the fetus at an early stage of development.

Formerly, scientists had assumed that the sickness early in pregnancy was related to a sharp rise in the hormone estrogen. But as some scientists point out, the sickness almost always occurs in the first three months of pregnancy, when organs are in a formative stage and therefore most vulnerable to trace poisons in the diet. Nausea would help a woman temporarily avoid foods that contain substances that are harmless to adults but that might cause birth defects or miscarriage. Studies have found that there is a higher rate of miscarriage among women who have little or no morning sickness.

A Bath for the Fetus

Q. What is amniotic fluid made of?

A. Fetal urine is the main component of the amniotic fluid that bathes the developing fetus. It is excreted into the fetal sac inside the womb as the fetus develops. Normally, the fetus "breathes" this fluid into its lungs, and it is essential for the normal development of the lungs.

The fluid contains some fetal cells, which can be drawn out in the process called amniocentesis and analyzed to find out whether the fetus has some genetic diseases.

Younger and Taller

Q. Has any statistical relationship been found between birth order and height?

A. Many studies have found a relationship between birth order and birth weight, with the later-born children in a family tending to be larger, and birth weight has some correlation with eventual height.

For example, a 1988 study at the Children's Hospital of the University of Kiel, Germany, investigated adult height in families with three or more adult children. The researchers found that the first-born siblings had a statistically significant deficit in average height as compared with the average of all the siblings, but the gap

amounted to only a fraction of an inch.

That study also found that men tended to increase in height with increasing birth order, while women did not, beyond the second born. The researchers concluded that birth order is a factor that contributes significantly to the variance of adult height among siblings.

The reasons for the tendency of younger siblings to be born larger are not clear. In an Israeli study that adjusted for a number of factors—like maternal age, education, marital status, religion, smoking, height and prepregnant weight, gestational age, and sex—birth weight increased with increasing birth order both in a large cross section of newborns and in a group of large families, where many of the possible variables would presumably be constant.

When X/Y > 1

Q. If genetically the ratio of human males to human females should be 1 to 1, why are there more females than males?

A. In fact, each year more males are born than females. The normal slight excess of male births is usually attributed to the slightly greater motility of sperm carrying the male, or Y, chromosome.

It was recently discovered that in the average ejaculation, there are slightly more sperm carrying the X chromosome, necessary for conception of a female, than there are sperm with the Y chromosome. But emerging studies of embryos that have been conceived through in vitro fertilization indicate that more males are actually conceived.

Scientists suggest that the Y-bearing sperm are so much better at fertilizing eggs than the X-bearing ones that even though there are fewer of them to begin with, they end up penetrating more than half of all eggs.

What happens to the sex ratio after conception is another story. For example, in societies where women are not valued as much as men, they may be purposely aborted, neglected, or actually killed. In others, where men are subjected to greater stress or risk (from wars, for example), more women survive longer.

The ratio of males to females can vary from time to time and from population to population, whether it is animals or people, and there are numerous suggested reasons. For example, it has been reported that women in lower socioeconomic groups, presumably under greater stress, bear slightly more girls than boys.

THE ADVANTAGES OF BEING FEMALE

Q. I recently read that infant girls usually have lower infant mortality than boys because of "biological advantages." What might these be?

A. The advantages for girls begin even before birth, and like so many other differences between the sexes may be tied to hormones. For example, significantly more male fetuses are spontaneously aborted or stillborn.

The reasons need more investigation, but they seem to

include sex differences in chromosomal structures and possibly a slower maturing of boys' lungs because of the effects of testosterone. In most societies this pattern of excess male mortality continues for the first six months.

An inborn biological advantage may persist at least until menopause because premenopausal hormones like estrogen offer some protection from heart disease caused by damage that narrows blood vessels.

The advantages may also include differences in susceptibilities and immunities to some disease-causing organisms. Still, some diseases are more likely to sicken and kill women.

Boy Twins, Girl Twins

Q. Are twins more likely to be one sex or the other? What about conjoined or Siamese twins?

A. Slightly more than half of all twins born are male. In conjoined or Siamese twins, the picture is complicated both by their extreme rarity and by their very low overall survival rate, which is variously given at 5 percent to 25 percent.

Conjoined twins are always identical, because they always come from a single fertilized egg and so are always the same sex. The developing embryo splits within two weeks of conception, but is only partly separated and continues to mature into two infants that are connected to a greater or lesser degree. Many die at birth and many have severe birth defects. About 35 percent survive just one day. However, historical studies of about 600 sets of conjoined twins over five centuries list more than 70 percent of the surviving pairs as female.

In the United States there has been a sharp rise in multiple births overall, associated with the two overlapping and related trends—older age at childbearing and increased used of fertility-enhancing therapies, many of which have multiple pregnancies as a side effect.

THE LONGEST DIGIT

Q. Why might the middle finger have evolved to be the longest?

A. There is no definitive answer, but its structure suggests a key role in the strong grip for hanging and power grasping.

The middle finger's length is one aspect of hand evolution that people have in common with apes. For many primates, the middle finger is the functional axis of the hand and the longest finger. All the muscles of the hand are arranged around an axis through the middle finger, so it is in position to exert the greatest amount of force, and the middle finger bones are the longest and most robust of any finger. All this suggests that this part of the hand blueprint came from apelike ancestors 5 million to 6 million years ago. Apes all have a powerful grasp, with well-developed fingers and a tiny thumb, unlike the long and very maneuverable human thumb.

The middle finger's bony anatomy suggests it is important in power grasping, while the thumb and index finger are for precision grasping, like holding a needle or stone tool. The third finger would be more devoted to the hook grip, used in hanging from a branch or, for modern humans, holding a heavy suitcase.

LIMITS TO GROWTH

Q. Will I grow up to be short like my mother or tall like my father?
A. Your adult height will probably be somewhere in between.

Some rough estimates can be made based on the growth patterns of many children. These methods assume that there is no disease or hormonal problem to hinder your genetic potential.

Because height at the age of 3 is about 60 percent of adult height for girls and 55 percent of adult height for boys, two times your height as a 2-year-old will be roughly your adult height.

Another rule of thumb uses your parents' heights. Add them, divide by two, then add three inches for a boy or subtract three inches for a girl. The result is said to be correct within two inches about 95 percent of the time.

Most often, however, doctors refer to standard growth charts. Your current height and age give an indication of where you fall relative to the average for children of your age and sex. Because you tend to stay in those same rankings as you grow, you can estimate your eventual height.

X rays of the gaps, or epiphyses, between the ends of

your bones do not predict eventual height but could determine if there is a condition that would prevent further growth. As you grow, the gaps shrink. When you complete puberty, the bones are fused, and no more growth takes place.

STROLLER POTATOES

Q. I see children old enough to walk being pushed around in strollers. Can this stunt development?

A. Experts suspect that the problem is not that tired parents are using strollers while shopping, but rather that they may be preventing their children from moving enough the rest of the time.

Child development specialists think that freely crawling, cruising, and toddling are a normal progression in a child's physical and cognitive development. But as soon as babies can move independently, they tend to be confined much of the time to strollers, infant seats, high chairs, playpens, and walkers. Young muscles can languish.

In 1997, research at Case Western Reserve University found that walkers, in which children can "walk" without seeing their feet, are likely to impair motor and mental development, apparently because movement is limited and children cannot freely explore their environment.

Many experts recommend more activity for very young

children, not necessarily with formal exercise but by allowing normal walking, running, and climbing—within the limits of safety—and playing games as simple as catch, tag, and chase.

THE PAINS OF CHILDHOOD

Q. My 4-year-old occasionally complains of excruciating but ill-defined nighttime pains in his calves and shins. Friends say they recall similar pains, which used to be called growing pains. What are they? What helps?
A. The description fits what doctors still call growing pains, or benign limb pains of childhood. The suffering is real, though the source is still not understood.

Most pediatricians do not believe growing pains are actually related to the child's growth. For one thing, the location is ill defined, not specifically at the limbs' growth plates, and for another, the pains occur mostly in the age group from 4 to 12, while the periods of most rapid growth are usually earlier or later.

About 20 percent of healthy children get growing pains, usually at night and usually in the legs. They are somewhat more common in girls than in boys. Once other possible causes, like injury or inflammatory disease, have been ruled out, such pains are diagnosed as growing pains.

Usually, the use of gentle massage, warm towels, and perhaps acetaminophen or ibuprofen is all that is needed to let the child go back to sleep. The pains have no lasting consequences, and the next morning some children do not even remember being awakened by the pains.

STRAIGHTENING THE SPINE

Q. I got a brace when I was 7 to correct idiopathic scoliosis. Is this still the standard practice? Is it true that this problem is disproportionately common in girls?

A. Idiopathic scoliosis, a sideways curve of the spine of unknown cause in adolescence, is now believed to require active treatment in only about 10 percent of cases. When treatment is called for, a cast or a device called a Milwaukee brace is widely used.

At least 60 percent to perhaps 80 percent of the cases are in girls, and girls tend to have worse curves and greater progression.

A review of studies on scoliosis and its treatment found that the Milwaukee brace, which has been in wide use since the late 1940s, has been the subject of few long-term studies to evaluate its results. Of the studies, most did not keep track of the progression of the curvature, making it uncertain whether patients with braces would have gotten worse without it.

Bracing tends to stop the curve's progression in 85 to 90 percent of patients, but there is wide variation in response. Most common is a moderate correction while the brace is worn with slow, steady progression back to the original curve after its use is tapered off.

Q. Why do so many teenagers have trouble waking up in the morning? Is there any way to train yourself to make getting up at 7 A.M. easier?

A. Morning sleepiness in teenagers (or in anyone else) may have a cause as simple as lack of sleep, perhaps because of too much homework after a part-time job, followed by television. In that case the cure may be to sleep more, cut down on some sleep-robbing activities, and gradually change bedtime and waking time in fifteen-minute increments until the person wakes refreshed at a desirable time.

However, some researchers suggest that there may be physiological causes for drowsiness in teenagers, not just in the morning but in afternoon classes, and some even say that teenagers have a natural sleep pattern that would make them go to bed at 2 A.M. and stay in bed until noon. They base these suggestions on studies of changing levels of the hormone melatonin, which is involved in sleep regulation, and on comparisons of the academic performance of teenagers with early and late school starting times. In one study, some students in an early starting group built up a sleep debt and suffered ten-second micronaps of drowsiness and lapsed attention in the afternoon.

CHAPTER 11

Our Bodies, Our Aches and Pains

MIX AND MISMATCH

Q. Why is your blood type—A, B, AB, or O—so important? I have never understood exactly what would happen if you got a transfusion of an incompatible type.
A. The most basic classification of the hundreds of ways to type human blood, the ABO system, is important because incompatible types of blood react with each other to form sticky globs of red blood cells that can cause serious and sometimes fatal blockages.

The classification was developed in 1901 by Karl Landsteiner, a Viennese-born pathologist and immunologist. For the first time, it explained why certain blood transfusions were beneficial while others caused severe problems.

The red blood cells of people with type A blood have a protein called the A antigen on their surface; people

with type B blood have the B antigen. In type O, the cells have neither A nor B antigens; in type AB, there are both antigens.

In one example of incompatibility, type A red blood cells (erythrocytes) have the A antigen, and the liquid part of the blood (the serum) contains antibodies against the type B cells; type B blood has the opposite configuration. So if the two types are mixed, the antibodies from the serum react to the "foreign" types, making the cells clump together.

MANUFACTURING BLOOD

Q. Will it ever be possible to synthesize blood the way insulin is made?

A. Artificial blood will not be available any time soon, and the need for donations will remain acute for the foreseeable future.

Synthesizing blood is much more complex than synthesizing insulin. Blood includes many different kinds of cells, while insulin is a protein.

There are ways to make hemoglobin, the protein in blood that carries oxygen, and most of the blood alternatives or substitutes that are being worked on are hemoglobin molecules aggregated together. They will serve as a bridge for people who need transfusions. However, the products that are being worked on would not be substitutes for transfused blood, except in emergencies or acute blood loss. Artificial blood cells are way off in the future, a decade or more.

Pigs' blood, with minor modification of the surface of its cells, would be compatible in humans, but its use raises other issues, like disease transmission, that would have to be worked out.

Many products have been available for some time that serve many, but not all, of the functions of plasma, the liquid that carries the cells and proteins of blood. These products can be used in certain situations.

THOSE ICY FINGERS

Q. When my fingers feel cold as ice indoors, how cold are they? What causes it?

A. What seems like frozen fingers is very subjective and could cover a wide range of temperatures.

Having fingers that are icy cold is a real sensation and can be very distressing, but it is usually not reflective of something dangerous. What is more dangerous is where one finger is cold or has changed color, or where feeling is absent, as it is in frostbite.

Skin temperature is controlled by the sympathetic nerves and the adrenal gland, part of the system for conserving or losing body heat. If it is cold outside, the system keeps heat inside the body by cutting skin circulation, reducing heat loss from the surface. If the nerves are overactive, the small blood vessels in deeper layers of the skin will contract, so the skin of the fingers may be cold while the body may not be.

Diseases that cause cold fingers are easily identified. They include Raynaud's syndrome; vasculitis, an inflammatory disease of the small blood vessels of the skin; and some arthritic diseases, such as rheumatoid arthritis. Smoking or caffeine can cause a problem if the sympathetic nerves are sensitive.

MEN, WOMEN, AND TEMPERATURE

Q. It seems that women are more sensitive to temperature changes than men. Is this true?

A. Sensitivity to temperature is much more complicated than a simple gender split, physiologists say. The reaction depends on many factors, such as exercise, previous conditioning, diseases, and any other tinkering with the complex system of signals to and from the hypothalamus that sets the body's temperature controls.

Some doctors suspect psychological factors make a difference, but it is difficult to separate psychological reactions based on a strong aversion to heat or cold from physiological responses. It is also possible that women do not dress as warmly.

But some male-female physiological differences may be significant in this area. Women tend to have more subcutaneous body fat than men, making them more likely to have a hard time dissipating body heat when temperatures rise. But women may do better than men as a room cools, because fat holds heat.

Larger animals have a smaller surface area relative to their volume than smaller animals, so they can stand colder temperatures with less heat loss. Men as a group are larger than women, although the size difference may not be enough to cause different reactions to temperature changes.

Some conditions, including an overactive thyroid, which is more common in women, can also cause hypersensitivity to heat.

The Big Chill

Q. If a body temperature of 107 degrees Fahrenheit is nearly always fatal, what low body temperature is generally fatal?

A. Medical literature contains few if any cases in which people have recovered after their inner, or core, temperatures dropped below 55 degrees Fahrenheit.

Core body temperature is usually measured by electronic sensors inserted into the digestive tract. It cannot be measured with an ordinary oral thermometer. Hypothermia is considered to exist whenever the core temperature is below 95 degrees.

Hypothermia occurs when the body loses heat faster than it can burn fuel to replace it. How severe a person's reactions to low temperatures will be depends on age and overall physical condition, with the very old and the very young the most vulnerable; about 80 percent of hypothermia deaths occur in those older than 65.

Age increases the risk for several reasons. The ability to generate body heat declines because the heart rate is slower, blood vessels do not contract as well, and the elderly often lose muscle tone (which generates heat) and body fat (which conserves heat and supplies body fuel).

Exposed skin, cold air, cold water, cold wet clothing, and sitting on a cold surface that conducts heat away from the body all greatly increase the rate of heat loss. Dehydration also adds to the risks.

The rates of chemical activity in the body are slowed by very low temperatures. Eventually the heartbeat and breathing may stop, but even a very slow heartbeat and rate of respiration may maintain vital body functions.

Immediate treatment for hypothermia victims involves stopping further loss of heat by wrapping the person in warm, dry blankets. Gentle handling is essential because jolts can cause fatally irregular heart rhythms.

LIVER SPOTS

Q. What causes liver spots and how can they be removed?

A. Liver spots, also called age spots or lentigines, are the result of hyperpigmentation, or the buildup of excess pigment in patches of the skin.

Liver spots have nothing to do with the liver and most often result from a lifetime of exposure to sunlight. Other possible causes include surgery, pregnancy, and some medications.

Liver spots are usually flat and brown, with rounded edges, and are most often found on the face, hands, back

and feet. If the spots have irregular borders, a dermatologist should look at them to make sure they are not melanoma, an invasive skin cancer.

As with so many other results of photoaging, age spots can be prevented or minimized by keeping out of tanning booths and direct sunlight at midday, routinely using sunscreen, covering arms and legs, and wearing a hat with a four-inch brim.

Treatment may involve bleaching or fading creams, which can take weeks or months; chemical peels; prescription drug treatments like tretinoin, a derivative of vitamin A, used with sunscreens; related over-the-counter products containing retinols; and surgical resurfacing, often with lasers, to remove the pigmented cells.

THOSE ACHING FEET

Q. Why do my feet get more tired and sore when I am standing still than when I am walking?

A. If we leave aside the possible presence of diseases like arthritis or diabetes, several mechanical factors could make standing still more tiring than walking. First, the action of the muscles in walking keeps the circulation of both blood and lymph going, preventing the pooling of fluids in feet and ankles.

For another thing, when a person stands still, each foot is supporting about half the body's weight all by itself all the time. But in walking, there is a momentary respite for each foot as the other makes contact with the ground. When the unburdened foot swings forward, it is free of weight-bearing duty for the time it takes to complete that stride.

Experts in easing the fatigue of those who must work standing up for long periods often suggest making a conscious effort to shift weight from one foot to the other and elevating the feet alternately on a footrest.

Another fatigue factor in standing still is that the surface underfoot remains the same. Studies of store employees and shoppers have found that there is less perceived fatigue even in walking if the flooring changes periodically, perhaps from wood to carpet. A resilient surface also helps.

Also, someone who plans to take a walk may have chosen more supportive and comfortable shoes than the person who unexpectedly encounters a long line at the bank.

THOSE ACHING EYES

Q. Why do eyes burn, itch, and often turn red and feel swollen after a night with too little sleep? Why the relief from a good night's sleep?
A. Such a condition is probably an inflammatory response to the irritation of having the eyes open and exposed to the drying air. The effect is common knowledge and a clear symptom of sleeplessness but has not been completely studied.

Eye lubricants and tears dry in the air and may form scratchy particles. The skin around the eyes is pretty loose, and it can gather fluid fairly easily, so the swelling can be a sign of irritation. Redness, a result of capillary dilation, is another sign of irritation.

In some cases, the condition might also be a manifestation of an allergy, especially one to mattress mites, microscopic arachnids that infest many people's bedding. In such cases, a water bed might help.

A night of sleep helps cure the condition by protecting the eyeball. Even if you can't sleep, keeping the eyes shut will help.

PROTECTING THE EAR

Q. What is the function of ear wax? Where does it come from?

A. Ear wax, known to scientists as cerumen (pronounced suh-ROO-mun), traps dust and dirt particles to keep them from going down the ear canal to the eardrum. Besides protecting against dirt and water and lubricating the canal, ear wax is slightly acidic, so some experts say it has mild antibacterial properties and helps fight the growth of fungus or bacteria that can cause the outer ear inflammation called swimmer's ear.

Ear wax is formed by small glands under the skin of the outer part of the ear canal. It builds up in different amounts in different people. It may get thicker and drier in old age. Normally, the ear is self-cleaning; the wax gradually dries up, flakes, and either falls out or can be gently wiped away.

If too much wax accumulates, the traditional warning never to use cotton swabs or other objects to remove it still holds true, because a small instrument can push the wax deeper into the ear canal, against the eardrum, where it can block hearing or injure the drum.

Your doctor can remove the wax or suggest commercial or homemade drops to soften it. In some cases, stronger prescription softeners can be called for. Softeners or even water should never be used without medical advice if a perforated eardrum is suspected, because an infection may result.

Not Just an Appendix

Q. Does the appendix have any use in the human body?
A. The vermiform (Latin for worm-shaped) appendix, a dead-end branch of the intestines, plays at least a minimal role in the immune system and is also sometimes regarded as a potential spare part for replacing things like a diseased urinary bladder.

Along with the adenoids, tonsils, and spleen, the appendix is classified as part of the mucosal immune system, a secondary set of immune organs along the mucus-lined membranes of the body. This system is also called the mucosal-associated lymphoid tissue, or MALT.

Like the rest of the gut, the appendix contains nodules called Peyer's patches where certain types of immune cells mature and "learn" whether or not to react to foreign substances by calling in cellular troops to attack the invaders. This function of the appendix is most pronounced in the first few decades of life and declines thereafter.

However, the appendix is not a specialized immune organ and is not essential to health. It is a vestigial part of the gut that was important in animals whose diets were composed of more fiber than humans consume.

It is not normally full of immune cells called lymphocytes, except when it becomes infected, and since it is a pouch, this occurs with some frequency. Then lymphocytes and macrophages, which clean up microbial invaders, infiltrate in order to deal with the infection.

WRINKLING OVER TIME

Q. Why do we wrinkle as we age? What can be done about it?

A. Skin changes that cause wrinkles do accompany aging, as the deeper layer of skin, the dermis, gets thinner. But it happens more quickly in sun-exposed areas and in people who smoke. The breakdown of two kinds of molecules, collagen and elastin, is at fault.

Collagen type 1 is the molecule that makes up the bulk of the skin. Loss of this type of collagen, the same type found in bones, affects the elderly, and smoking can make it worse.

At least five studies have found that smoking is associated with "smoker's face," one that is prematurely aged by fine wrinkles that can accentuate the coarser wrinkles that occur along the lines of expression.

As for elastin, the stretchy molecules that help support the skin, ultraviolet rays cause direct damage, breaking the molecules down.

To help avoid wrinkles as you age, avoid smoking, stay out of the sun, routinely use hats and sunscreens, and use a good moisturizer, which holds water in the dermis and plumps it up.

LOSING COLOR

Q. Is there any treatment for vitiligo, the loss of skin pigment that leaves white patches on the body? And what causes it?

A. There are treatments for vitiligo that are at least partly effective in many cases, but its cause is still under study.

A prominent theory is that an autoimmune reaction is involved. The condition may also be associated with certain ailments, including thyroid disorders, and may flare up after injury, stress, or severe sunburn.

Patches occur in all races when pigment cells in the skin, called melanocytes, are destroyed and can no longer produce the pigment melanin. The most common treatment involves a drug called psoralen, taken orally. When followed by exposure to natural sunlight or ultraviolet light, the drug seems to make pigment cells migrate from hair follicles to white areas.

In some patients, steroids stop the spread of pigment loss. In others, such as the singer Michael Jackson, doctors use chemicals to lighten unaffected skin areas so they blend in with those where pigment has been lost.

Medical authorities also recommend avoiding excessive exposure to the sun. Tanning darkens normal areas, which makes the light patches more prominent because they do not tan.

SOAKING UP THE SUN

Q. How do people get a tan?

A. Exposure to the sun's ultraviolet radiation darkens granules of the pigment melanin in the surface layers of the skin. Part of the radiation also stimulates pigment-producing cells, called melanocytes, in deeper layers of the skin, causing a delayed-reaction tan, about three days after exposure.

When the top layers of skin wear off, so does the tan.

All adults have about 60,000 melanocytes per square inch of skin. Skin color is determined by how much pigment the cells make and what color it is; it ranges from black to light tan. Other factors that stimulate melanin production include X rays, heat, certain chemicals, hormones, and drugs (usually in combination with sun exposure), and inflammatory skin diseases.

Dermatologists warn that tanning goes hand in hand with skin aging and a higher risk of skin cancer. Ultraviolet B, or UVB, radiation, the so-called burning rays, is the kind most sun products protect against. They usually offer little protection against the ultraviolet A, or UVA rays, the so-called tanning rays. These are the rays that affect deeper layers of the skin, causing the breakdown of supporting tissue, resulting in premature aging, wrinkling, and a leathery appearance.

Some newer sunblocks protect against both kinds of radiation, but nothing works as well as avoiding midday sun exposure altogether and wearing a hat and long sleeves.

DAMAGING THE DNA

Q. How does ultraviolet radiation cause skin cancer?
A. The prevailing theory involves damage to the DNA of skin cells. It is the shorter UVB rays, which penetrate only the top layers of the skin, that are suspect, while the

longer and more deeply penetrating UVA rays cause wrinkles and aging.

In tissue cultures, UV damages the DNA of cells, but humans have an enzyme that repairs it. Due to a genetic defect, some people lack the repair enzyme. The hypothesis is that in the body such DNA damage occurs all the time and is constantly repaired, but some cells do not get repaired, or get repaired improperly, and this is how skin cancer begins.

Researchers found a specific kind of DNA damage in a gene called CQ that occurs in this way. In the error, two DNA units of the type designated as thymine are side by side, instead of two units of the cytosine type. The error is called a thymine dimer. It is presumed that the brakes on cell multiplication come off because of it, leading to uncontrolled proliferation of cells into a tumor.

Ultraviolet radiation is implicated in the vast majority of nonmelanoma skin cancers, like basal cell carcinoma and squamous cell carcinoma. It is also linked to melanoma, though less clearly.

PUTTING THE PINCH ON A NERVE

Q. What is a pinched nerve, and what is it pinched between?

A. A pinched, compressed, or entrapped nerve can be surrounded by swollen tissues or impinged upon by bony processes in many areas of the body. The cause can be injury, disease, or even pregnancy. The result is pain, numbness, tingling, and weakness in the part of the body to which the nerve normally carries impulses.

One kind of nerve entrapment, carpal tunnel syndrome, is produced in the wrist because of repetitive stress from movements like typing; the resulting swollen tissues compress the median nerve, which runs through a bony tunnel in the wrist called the carpal tunnel. Similar problems can affect the ulnar nerve, which runs through the elbow, and another nerve in the arm, the radial nerve.

The pain of sciatica results from compression of the sciatic nerve, which runs down through the buttocks to the back of the leg. The cause is often a herniated spinal disk; normally, this gelatinous cushion with a tougher outer ring separates the vertebrae, but injury can squeeze out the soft core so it puts pressure on a nerve.

A pinched tarsal nerve or plantar nerve can cause numbness in the foot. Some other commonly compressed nerves include the femoral nerve, which extends from pelvis to knee, and the peroneal nerve, along the side of the leg.

RUNAWAY SCARS

Q. What is a keloid and how do you get one?
A. A keloid is a scar that does not know when to stop forming, becoming large, shiny, smooth, and often pink and dome-shaped. It is not known why some people get overgrown scars after injuries, surgery, or acne, but

keloids are more common among people of black and Asian descent, so a genetic factor is suspected.

In normal scarring, after the inflammation that follows an injury subsides, scar tissue begins to form, along with tiny new blood vessels. Cells in the skin around the injury, called fibroblasts, produce collagen, a fibrous connective tissue. As more and more of the fibers link up, the scar becomes harder.

In a keloid, the process continues long after the wound is covered over, and the scar can become quite large. Keloids are not dangerous but can be disfiguring, tender, and sometimes itchy.

Removal of a keloid by surgery or the use of lasers, followed by corticosteroid injections at the site, is sometimes but not always successful, and can cause even worse scarring.

Someone with a tendency to form keloids may want to avoid plastic surgery, though doctors can sometimes use hidden incisions in facial surgery or avoid making cuts in the periphery of the face, where keloids are more likely to form.

TOO MUCH OF A GOOD THING

Q. Can a human breathe pure oxygen?

A. Yes, but it is not a good idea except under very special circumstances, because too high a concentration of oxygen or oxygen at too high a pressure can be toxic. For example, when a diver gets too much oxygen at high pressure, oxygen toxicity can cause lung damage and even convulsions.

However, pure oxygen in the form of O_2 is sometimes administered in an entirely closed chamber at greater than the normal air pressure at sea level; this is called hyperbaric oxygen therapy. Hyperbaric oxygen may be called for in cases of carbon monoxide poisoning, decompression sickness in divers who ascend too quickly, gas gangrene, smoke inhalation, some burns and infections, skin grafts that are about to fail, and some other conditions. In such cases, the extra oxygen may be lifesaving, but even then, it is usually used in repeated spurts.

Another well-known adverse effect of too much oxygen is the blindness in premature infants that is called retinopathy of prematurity, or retrolental fibroplasia. The blood vessels that supply the retina of a prematurely born baby are not fully developed, and in the presence of high concentrations of oxygen in an incubator, they may continue to grow in an abnormal pattern, destroying vision. Too much oxygen can also cause lung damage in infants.

OLDIES BUT GOODIES

Q. As a reasonably healthy 68-year-old, I have wondered, Is there an age limit for usable donations of organs like the heart or liver?
A. Organs may come from newborns on up, with no specific age limit. There was more of an age limit in the past, but now it is decided on a case-by-case basis.

The oldest known donor of a liver was 92, in Canada, and there was one who was at least 85 in the United States.

At the time of a donor's death, medical professionals determine whether organs are usable. Needed organs include the heart, kidneys, pancreas, lungs, liver, and intestines.

There is a general age limit of 70 for donation of tissue, including the eyes, skin, bone, heart valves, and tendons.

An important factor in donation acceptance is prompt recovery. Telling family members now, not in a will or letter, that you want to be an organ and tissue donor is the best way to make sure that your wishes are carried out.

DEATH BY AIR BUBBLE

Q. Is it true that air is released from a hypodermic syringe because an injected bubble could kill you?

A. Crime novels to the contrary, it is hard to dispatch someone with a needle unless you are expertly trying.

It is standard practice to make sure air is cleared from a syringe before pushing the plunger, but the chances of death occurring because of the air a fine needle might carry in a normal injection into muscle or skin are rather low; a small bubble would normally be dispersed in the tissue.

A bubble of air in the circulatory system, called an air embolism, can potentially be fatal, but usually only if the air is introduced into a fairly large vessel. If an arterial bubble cuts off oxygenated blood going to the heart muscle or a key part of the brain, for example, it could cause a heart attack or stroke. An air embolism in a large vein can also cause a dangerous air lock in the heart.

The main risk, however, lies in procedures like intravenous drips or in surgery. The rule in open heart surgery is "no air, anywhere," and heart-lung machines have filters for bubbles. There are also detectors for intravenous procedures.

FINAL CHANGES

Q. How long does it take for rigor mortis to set in? Is it a good indicator of the time of death?

A. Rigor mortis is gradual, usually noticeable about three to four hours after death, and the timing varies with several factors, including the person's activity just before death and the temperature of the body's surroundings. How long it takes the muscle stiffness to subside also varies, with rigidity usually reaching a peak after twelve hours and nearly disappearing in twenty-four to thirty-six hours.

Rigor is not a good indicator of the time of death. Coroners use the body's stiffness only in combination with other factors in estimating when a death probably took place.

Rigor mortis, literally the stiffness of death, comes about because of chemical processes within muscles. Their normal operation, in which filaments slide past each other in contraction or relaxation, depends on energy from molecules of adenosine triphosphate, or

ATP. As the supply of ATP is depleted after death, two important muscle proteins, actin and myosin, become chemically locked. Eventually, decay begins and the muscles soften.

Two other postmortem changes used in estimating time of death are algor mortis, or cooling, which proceeds at about 1.5 degrees Fahrenheit an hour under normal conditions, and livor mortis, the pooling and settling of the blood, which starts immediately after death and becomes fixed in about twelve hours.

CHAPTER 12

Things You've Always Wanted to Know

An Icy Stare

Q. Why don't your eyes freeze when you are skiing or walking the dog in subzero temperatures?

A. The eyes have a very efficient radiator as well as a surface deicing system, at least in theory. The eye has a system of blood vessels that provides a constant irrigation of warm blood through and around the eyeball to keep it warm. This makes freezing very unlikely.

As for the surface of the eye, tears have a high salt concentration, which keep them from freezing at 32 degrees, as plain water would, and the eye is constantly coated with tears.

It is not impossible to freeze the eye, but it requires very low temperatures. For example, a freezing procedure called retinal cryotherapy, using a probe with liquid nitrogen, has been used to destroy certain parts of the retina, and the same procedure is also used to kill the

cells of certain cancerous growths.

In the procedure, part of the eye is frozen for several seconds, and a visible ice ball forms, but the eye immediately defrosts when the probe is removed.

Sneezing After Hours

Q. Why don't you sneeze when you are asleep?

A. Nobody has specifically studied this aspect of sneezing, but there are logical reasons that would explain why people do not sneeze overnight.

The reason people sneeze is that the inside of the nose swells up, much like blowing up a balloon. When a windblown particle of anything, an allergen or a particle of dust, hits the swollen lining, a sneeze results. When the wind blows or a door opens or a strong odor wafts in, a sneeze is likely to be set off.

In most bedrooms, there is no airflow at night, not even anyone walking around stirring up dust, so no particles blast the nose. Many people complain of sneezing attacks when they first wake up, throw off the covers, open closets, and stir up particles that settled overnight.

The nose lining actually tends to be even more swollen when people are lying down, but the lack of a breeze protects against a sneeze. Even with air-conditioning there is generally a filter, so particles get stirred up only if the unit is turned off for a while in a dusty room. In that case, the first blast of air may throw particles at the nose.

An allergic person by a window in pollen season will sneeze all night. But in that case, the person would probably not be asleep.

DANGEROUS HICCUPS

Q. Can hiccups kill you?

A. Ordinary hiccups, resulting from things like overeating and stress, are usually benign. But hiccups can indicate a serious problem, and a prolonged, uncontrollable bout may lead to debilitating consequences like fatigue, weight loss, depression, problems with heart rhythm, esophageal reflux, and possibly exhaustion and death in a weakened patient.

In a famous case, Pope Pius XII had long periods of hiccups, associated with gastritis, though he eventually died of a stroke.

A hiccup is a spasm of the diaphragm, the sheet of muscle that controls breathing, accompanied by a sudden intake of breath and the closing of the epiglottis, the bit of tissue at the back of the throat that can close the airway.

The underlying cause of troublesome hiccups should be investigated. For example, the complicated nervous pathways involved in hiccups can be affected by diseases like multiple sclerosis; in such cases, and for other serious hiccups, muscle relaxants are often prescribed.

Other causes include gastrointestinal problems such as obstruction or inflammation, complications of AIDS or side effects of its treatment, nervous system tumors, encephalitis and meningitis, alcoholism, and problems with many medications.

PLEASURE AND PAIN

Q. Why does eating something really cold, like ice cream, sometimes give me a headache?

A. You are not alone, because about 30 percent of the population seems to have the same problem. The classic ice cream headache is a stabbing pain that usually peaks after thirty to sixty seconds, though it can last up to five minutes.

A doctor who gets such headaches, Dr. Robert Smith, founder of the Cincinnati Headache Center at the

University of Cincinnati College of Medicine, experimented with crushed ice applied to different parts of his mouth to see what set off the pains. The culprit, his research indicates, is at the back of the palate, from which a mass of nerves called the sphenopalatine ganglion stretches into the head. The nerves control the dilation and contraction of blood vessels, and dilated blood vessels in the head are known to cause several kinds of headaches.

Dr. Smith also found that he was able to induce such headaches only on warm days, though the internal mouth temperature remains constant. Other researchers have suggested that this is because in hot weather, a person tends to gulp cold drinks or wolf down lemonade so that the cold stuff is more likely to hit the palate hard. Ways to avoid the pangs might be to let the ice cream warm up in the front of the mouth before swallowing or to swallow it in such a way that it does not linger on the palate.

TELEVISION VISION

Q. My mother says sitting too close to the TV will ruin my eyes. Is she right?

A. Not according to experts. In fact, children can focus up close without eyestrain better than adults. That is because the lens, which changes its focus for nearby and faraway viewing, tends to lose some of its flexibility as people grow older.

Usually, sitting very close to the television set is a habit that children grow out of. However, sitting close to the screen to get a clearer image may be a symptom of uncorrected nearsightedness, and children who persist in doing it should be tested for myopia.

A lot of the traditional parental warnings about ruining your sight are only partly valid, according to ophthalmologists. For example, reading in dim light may tire your eyes, but it will not damage them.

Eye fatigue is a risk when reading in dim light and when using a computer screen for long periods, but it is reversible and preventable with commonsense measures such as accurately ground corrective lenses designed for the distance at which they will be used and sufficient blinking to lubricate dry eyes. Regular breaks to look up or across the room are strongly recommended, because looking at objects farther away usually relieves the strain on eyes.

If vision blurs or eyes tire very easily, it is time to get an eye examination from an ophthalmologist.

Ticklish Subject

Q. Why can't you tickle yourself?

A. The brain can tell which tickling sensations are caused by one's own actions and gives them low priority, so that it can be more receptive to sensations from outside sources that may be more urgent. One study suggests how brain areas may interact to do this.

In the study, a magnetic resonance imaging device scanned brain activity while volunteers had a palm tickled either by themselves or by the experimenters, using a tickling machine, a box with a pivoting rod. A piece of soft foam at the end of the rod could be moved by either the subject or an unseen experimenter.

The researchers found that the areas of the brain that process touch information, the somatosensory cortex, were more active when subjects were tickled by someone else than when they tickled themselves. The cerebellum was less active when the subject was making a movement that resulted in a touch sensation than when the movement did not produce a sensation.

This suggests that the cerebellum anticipates the sensation caused by the movement made by subjects tickling themselves and sends a message to the somatosensory cortex telling it to decrease its activity, so that the sensation is dulled.

A Pain in the Side

Q. What is a stitch in your side? What can you do to relieve the pain?

A. Many of the explanations offered for the sharp pain runners feel during hard breathing, usually on the right side, involve muscle stress or spasms in or near the diaphragm.

Because the muscle stress is probably caused by a rhythmic, repetitive activity, the solution is to develop techniques to break up that stressful rhythm. Runners could try changing their breathing patterns by breathing slower and deeper and forcefully exhaling through pursed lips. Other suggestions involve changing running position by leaning forward, backward, or sideways, stretching before running, having the stomach as empty as possible before running, and changing the foot one lands on when exhaling.

A more direct explanation is that the rhythm of running causes bouncing of the liver, stretching the ligaments that link liver and diaphragm. In this case, the suggestion is to stop running and use a hand to press the liver up against the diaphragm until the pain ends.

Catching Up on Sleep

Q. Can you "pay back" a sleep debt?

A. Yes, eventually, but it is not a simple matter of sleeping longer on a weekend after being denied adequate sleep for some days. In extreme cases, experts say, you may need six weeks of adequate sleep to repay a long-term sleep deficit.

Several studies have found that the amount of recovery sleep required to restore performance (though not mood) increases the longer someone has been awake. It might take two hours of sleep after one sleepless night, five hours after two nights.

Daytime napping can also help repay a sleep debt, and some experts even suggest that a two-hour "prophylactic nap" early in a long, planned period of sleep deprivation can help maintain productivity.

What "adequate sleep" means varies from individual to individual, and many do not even realize they are in debt. One test is to see how long you sleep before waking up naturally, feeling refreshed, with no alarm.

Sleep requirements hinge on many factors, including how efficient the sleeper is in getting the right kind of brain waves at the right time during sleep; scientists do not fully understand the function of the different kinds, but slow-wave sleep seems to be the most restorative after a sleep deficit. Most scientists seem to think that about half your normal sleep is needed to function the next day and that most people can get by on five hours, but as the debt grows, they may find themselves drowsy in meetings—or behind the wheel.

WHAT'S THAT?

Q. What does that little thing hanging at the back of our throat do? My science teacher said it does nothing.
A. The small piece of soft tissue dangling over the tongue is the uvula. It is variously described as U-shaped or tear-shaped, or classically speaking, grape-shaped; its name comes from the Latin word for grape, *uva*.

While some anatomy experts think it is only a vestigial organ, not all would agree with your teacher that it has no function. Some might say that it helps keep grapes (or anything else) from going down your breathing passage when you swallow, and that is the action that many standard anatomy texts describe.

The uvula has its own little muscle, the musculus uvulae, to help it stiffen and change shape, so it helps fill in the space at the back of the throat.

Singers credit the uvula with letting them produce a vibrato, or wavy up-and-down sound, and the jazz star Anita O'Day says that her vibrato-less vocal style, using many separate notes instead, arose because she lost her uvula in a tonsillectomy.

The uvula is one of the soft-tissue structures that is often implicated in snoring and sleep apnea, or interrupted breathing during slumber. Some treatments for these conditions involve removing excess flesh from the uvula and surrounding areas. The surgery, called uvulopalatopharyngoplasty, tightens up flabby tissues to enlarge the air passages.

LUMP IN THE THROAT

Q. What causes you to get a lump in your throat?
A. If the lump is stress related and does not involve serious difficulty in swallowing, it is probably just an exag-

gerated awareness of the epiglottis or spasms of muscles surrounding the throat.

After ruling out physical ailments that can cause the feeling of a lump in the throat, like throat cancer or esophageal reflux, in which some stomach contents are released upward, doctors call the condition globus hystericus.

Globus hystericus can be a small component of a larger set of symptoms that psychiatrists call generalized anxiety disorder, which may require psychiatric treatment or anti-anxiety drugs. But for such a diagnosis, the sufferer would have multiple, long-lasting symptoms from several of these groups: motor tension (jumpiness, shakiness); hyperactivity of the autonomic nervous system (sweating, pounding heart, clammy hands, dry mouth, the lump in the throat); apprehensive expectation (fear, rumination, anticipation of misfortune) and vigilance and scanning, with results like distractibility or insomnia.

If you can identify the stress that leads to the lump, and it is something as simple as a sad movie or a glimpse of an old boyfriend or girlfriend, you can probably avoid the sensation by avoiding the source of the stress or finding a better way to deal with stress, like meditation, breathing exercises, or yoga.

NERVES AND TEETH

Q. Having recently had a root canal procedure, I've been wondering: Why do teeth contain nerves?
A. Like other important organs, teeth have nerves for sensation and protection.

The diseased "nerve" removed in a root canal procedure is the common term for the combination of nerves and blood vessels that form pulp, the soft interior of the tooth. There is also a bundle of nerve fibers in the ligament that attaches the tooth to the bone. Together, these nerves let the teeth integrate with muscles, the brain, and the temporomandibular joint, responsible for the up-and-down motion of the jaw, in the complex activity of chewing.

The ability to know where the teeth are at all times is called proprioception, and nerve sensation is necessary to guide the teeth. If your bite is not quite right after dental work, or if your teeth shift, you can sense it immediately.

The nerves also let the teeth perceive heat, cold, and pain and protect them from harm. These sensations can warn of decay. Often, if there is a large cavity, the tooth becomes sensitive to sweets, as part of the body's defense mechanism.

No Stomach

Q. Can a person live without a stomach?
A. The stomach is not essential for human survival. Many people have survived and adjusted to total or partial surgical removal of the stomach because of diseases like stomach cancer.

The stomach is essentially a reservoir that allows people to eat the quantity of food they want, emulsify it, and pass it gradually into the small intestine. Without a stomach, one cannot consume the quantity of food one was able to before and usually must have more frequent and smaller meals.

There is an initial uncomfortable transition period after the loss of the stomach that may last months before the person can adjust to the loss.

The operation to remove a stomach is called a gastrectomy. It can be either partial or total. Doctors often place an external feeding tube directly into the small intestine, a procedure called a jejunostomy, to provide consistent nutrition during recovery and acclimatization.

If a person without a stomach eats too much, it is very quickly apparent because of an uncomfortable feeling of being very full. Other possible problems include diarrhea, a condition called dumping syndrome, in which nutrients are moved too rapidly into the small bowel, and upward reflux of bowel contents.

BRAIN FOLDS

Q. What do brain wrinkles have to do with how smart you are?

A. Brain wrinkles seem to have more to do with what makes humans smarter than lower animals than with what might have made Einstein smarter than you. In normal human brains, large grooves called fissures, small ones called sulci, and outward folds called gyri follow a standard plan from person to person.

Lower animals have smoother brains, apparently for evolutionary reasons. Cognitive brain activity takes place in cells in the thin surface layers of the cerebral cortex, and skull size did not increase as fast as brain complexity did as the species evolved. Convoluted human brains pack a larger surface area into a small container.

Human intelligence appears to be related to the branching of brain cells and the formation of complex links between them, not the shape of the platform where the links take place.

Only in a rare disease called lissencephaly ("smooth brain") is a lack of complex brain folding linked to mental deficiency. The condition is caused by the failure of developing nerve cells to migrate normally.

As for Einstein, recent studies of his preserved brain suggest that his intelligence may actually have been related in part to the absence of a fissure in a brain area involved in mathematical thinking, the inferior parietal lobe. The theory is that more links can be established between its cells because the groove is missing.

PERFECT PITCH

Q. What is perfect pitch, and why do some people have it?

A. Perfect pitch, which is also called absolute pitch, is the ability to name any note (such as A, B flat, C sharp) just from hearing the tone, without reference to other notes in a scale. Relative pitch, which is far more common, is the ability to tell what note is being played by judging the interval between its pitch and that of a reference note.

Each note in the scales used in most music is a sound resulting from a particular number of vibrations per second; for example, most modern pianos are tuned so that the A above middle C sounds at 440 vibrations a second.

How perfect pitch is developed or why some are born with it has been the subject of much study recently, and strong evidence points to both genetics and early musical training.

In a 1998 study of 600 musicians by researchers at the University of California at San Francisco, nearly all of those who had perfect pitch said they had started their training by age 6. The younger the musicians were when they started lessons, the more likely they were to have perfect pitch. Those with perfect pitch were also far more likely to have family members with the same ability.

Brain activity in response to music is now being closely studied by techniques including PET scans, and researchers have found that the area of the brain involved in processing sounds, the auditory cortex, develops clusters of specialized brain cells that are more active in response to piano tones in skilled musicians than in others.

While that study found no difference in brain activity in those with perfect pitch, other studies have found a sound-processing region in the left brain, part of the planum temporale, that is relatively larger than the same area in the right brain in those with perfect pitch and is active when they try to name a note.

This difference was found, the researchers said, even though most music processing is done in the right-hand side of the brain. One theory is that the overall contour and melodic functions of music may be handled by the brain's right side, while the task of naming notes and identifying their pitch could be concentrated on the left.

RING AROUND THE COLLAR

Q. Why does sweat leave a yellowish stain?

A. The most likely culprits for yellow clothing stains are body secretions called apocrine sweat and sebum, the oily secretion of the sebaceous glands. Deodorants and antiperspirants may also play a role.

The sebaceous glands are usually associated with hair follicles. Cells filled with fatty droplets die and burst, providing lubrication for the skin and hair. When the oils are exposed to air, they oxidize, turning yellowish, and if not quickly removed by thorough laundering, they can permanently yellow clothing. Sebaceous glands at the back of the neck cause "ring around the collar."

The underarms and groin are rich in apocrine sweat glands, which produce secretions that are milky and usually odorless until acted on by bacteria. Apocrine sweat contains many chemicals, including the acidic substances that produce the characteristic underarm odor.

The most copious human sweat, the kind produced by

widely distributed sweat glands called eccrine glands, is probably not a big stain producer unless the skin or clothing is already soiled. This kind of sweat is the body's chief means of temperature regulation, providing for cooling by evaporation. It is about 99 percent water, with only about 1 percent of it consisting of salt, urea, ammonia, and uric acid.

CHAPTER 13

Eat, Drink, and Be Healthy

ALL PIZZA, ALL THE TIME

Q. Can a person survive by eating only slices of New York City pizza?

A. Yes, if it is a real cheese pizza with real tomato sauce.

For a complete diet, first you need to consume a sufficient quantity of food to get enough calories. Then you need a good source of protein and a supply of certain essential nutrients, like vitamins B_{12} and C, found only in certain kinds of foods. Other vitamins, minerals, and useful nutrients like antioxidants should also be present in some amount.

Pizza mixes a lot of ingredients and can potentially meet all these requirements, as long as you eat enough slices.

First, pizza has plenty of calories, as dieters know. Protein is found in both the wheat crust and the cheese. Vitamin B_{12}, which is found only in animal products, is supplied by the cheese, a dairy product. (Beware of imitations; real pizza cheese is stringy when hot.) Vitamin C is found only in vegetables, the fresher the better, but canned tomatoes or tomato sauce will do. It would be preferable to include some slices of fresh tomato. A few greens on top, like parsley or broccoli or spinach, wouldn't hurt.

Vitamin E is found in the olive oil that good pizzas are made with. Vitamin A comes from the tomatoes, vitamin D can come from the sun, and vitamin K you make yourself. To top it off, tomato sauce is a good source of nutrients like lycopenes, with their rich antioxidant potential.

All pizza is not a perfect diet, but potentially you could survive for quite a long time.

PEANUT: FRIEND OR FOE?

Q. Is peanut butter good for you or bad for you?

A. It can be either very healthful or very noxious, or both. A study by scientists at the U.S. Department of Agriculture, done with money from the Peanut Institute, suggests that peanuts may contain enough resveratrol, the compound in red wine associated with a low rate of heart disease, to be helpful to human health.

The levels per ounce the scientists found were only about half those in wine, and the average serving of peanuts is only an ounce, not a whole wineglass.

And as George Washington Carver found, the peanut is very nourishing, perhaps too nourishing for weight watchers. It is rich in proteins, carbohydrates, minerals, and some vitamins. It contains no cholesterol, being a plant food, not an animal food, but it is full of oily calories; just a tablespoon of crunchy peanut butter has about 94 calories.

Another problem with peanut butter is that people who are allergic to peanuts tend to be severely allergic, subject to anaphylactic shock, which can stop breathing and cause death, if they eat even a trace. An undetectable amount of peanut butter used as a thickener in a prize chili recipe killed a Rhode Island student in 1986, leading to warnings about peanut butter hidden in unexpected dishes. A peanut allergy may be accompanied by severe allergies to other legumes.

THE TROUBLE WITH MOLD

Q. How risky is aflatoxin, a toxic carcinogen that can grow on peanuts, for American consumers?

A. The short answer is that aflatoxin is of very little concern. Aflatoxin is produced by the fungus *Aspergillus flavus*, and most people don't eat moldy food, because it tastes bad. Documented human aflatoxin problems usually occur in tropical countries, where the wet climate is conducive to mold growth, or in extremely poor ones, where people are forced to eat any food they can get.

Second, humans are much less affected by aflatoxin than animals, especially some birds. Even someone who tried to commit suicide with aflatoxin, a lab worker, did not die or suffer much long-term harm. It causes liver cancer in animals, but a big study of diet in China found no link between it and human liver cancer, except in those who had had hepatitis B.

Third, the U.S. Department of Agriculture works

with all big farmers to monitor all crops, such as grains and peanuts, that are susceptible to mold. The USDA designed a machine that kicks out suspicious peanuts, and even if you get a moldy one, chances are it is a different kind of mold. As for peanut butter, big producers test both raw materials and finished batches for aflatoxin.

TEENAGERS AND CHOCOLATE

Q. Does eating chocolate cause acne or make it worse?
A. Repeated studies have failed to find any link between eating chocolate (or anything else) and teenage breakouts. As long ago as 1950, a study by a dermatologist compared the results of giving teenagers with acne candy bars that just tasted like chocolate and those made with real chocolate. Photographs showed no difference in new breakouts.

In another study, 65 acne sufferers ate large amounts of chocolate daily; 46 showed no change in their acne, 10 got better, and 9 got worse. When the same patients received look-alike candy containing no chocolate, 53 showed no change, 5 got better, and 7 got worse.

In still another study, of Navy midshipmen with acne, four weeks of eating at least three bars of chocolate daily had no effect on skin lesions.

Some doctors suggest that an individual who observes breakouts after eating certain foods should avoid those foods, but allergists say chocolate is rarely if ever a factor in food allergies.

Some people fear the caffeine content of chocolate will keep them awake, but it is minimal, chocolate makers say. An ounce of milk chocolate has 5 milligrams of caffeine, an ounce of bittersweet chocolate has 5 to 10 milligrams and a 6-ounce cup of cocoa has about 10 milligrams, as against 100 to 150 milligrams in a cup of brewed coffee.

FEEDING THE NAILS

Q. Does eating gelatin really make your nails stronger?
A. Not any more than eating enough of any other protein. If someone is severely protein deficient, high quality protein like that in gelatin might make a difference, but the average person gets plenty of protein.

It's the same with calcium. It would help your nails if you were totally deficient, but if your bones are falling apart, you don't care about your nails.

To make a real difference in nail strength, treat your nails the way you treat your skin. If you moisturize your hands after you wash the dishes, for example, rub the product into the nails as well. Keep your nails out of harsh chemicals. Wear gloves to do housework.

Someone whose nails are really brittle might even wear latex gloves when washing his or her hair. Avoid nail polish and other drying nail-care products. Polish covers nail abnormality but actually makes it worse. Polish remover, even the kind without acetone, is very drying and hard on nails.

What if your nails split rather than break? Longitudinal splitting is something that occurs with age. Ridges and valleys develop, and splitting occurs along fracture lines, because the nails are drying.

Again, help for this problem lies in moisturizing. Wet the nails and cover them with Vaseline or even alpha hydroxy acid. It is not a quick fix, but must be done until the nail grows out entirely, which takes six to eight months.

The other kind of splitting, called onychoschizia, which occurs in layers at the tip of the nail, is something seen most often in chronic polishers or people who use fingernails as tools.

JUST ADD WATER

Q. I'm confused by what health authorities say when they recommend drinking eight glasses of water a day. How big is a glass?
A. They are usually talking about six to eight glasses of 8 ounces each, for a total of 64 ounces a day.

It does not hurt to drink more, especially if you are exerting yourself. You might weigh yourself before and after a workout and drink 16 ounces of water for every pound you lose.

With a loss of just about 2½ quarts, you start to have serious problems, like skin shrinkage, muscle weakness and fatigue, irritability, dizziness, and headaches.

Another time to drink extra water is when you are

on a diet to lose weight, to help the kidneys flush out the waste products from the extra flesh you burn, and to prevent constipation. Try drinking a glass before and a glass during each meal to control food intake. And at high altitudes, drink up to twenty glasses a day.

What nutritional authorities are talking about is not coffee, soda, or beer, but hydrating liquid, preferably plain water. Other beverages may contain diuretics (especially alcohol and caffeine) that make you lose water, on balance.

And don't count on your thirst to remind you when it is time to drink. The body can easily lose a quart or more of water before your thirst may signal you, and when you do take a drink, the feeling of thirst goes away long before you have actually replenished your water supply.

CALCIUM WITHOUT COWS

Q. I read that quinoa, a grain from the Andes, is very high in calcium. How high? What are some other nondairy sources of calcium?

A. Quinoa is a significant source of calcium, but you would have to eat a large amount of it to get a full daily supply. The National Academy of Sciences recommends 1,000 milligrams of calcium a day for most people, more for those over 50, and according to the United States Department of Agriculture, a cup (eight ounces) of cooked *Chenopodium quinoa* provides about 102 milligrams. That is more than some other grains, but a cup of cow's milk has about 300 milligrams.

There are other nondairy sources of calcium that are richer in calcium. Just two corn tortillas would provide about 120 milligrams, for example. Many greens are especially rich in calcium, and a cup of cooked turnip greens might offer 450 milligrams. Canned sardines and salmon with the bones in are good sources, too, as are some beans, peas, and seeds.

However, quinoa is a generally nutritious food, rich in protein, and it has several vitamins and minerals in it, including folate (49 micrograms a cup), a vitamin associated with a greatly reduced risk of birth defects when eaten by pregnant women.

ACID UPON ACID

Q. Is taking aspirin with orange juice a no-no? What about alcohol?

A. The *Physicians' Desk Reference for Nonprescription Drugs* (PDR) does not advise against taking normal doses of aspirin in addition to acidic foods or alcohol, itself a powerful drug, but there may be deleterious interactions for some people.

That is because aspirin is an acid, acetylsalicylic acid, and when aspirin is taken with acidic foods and drinks, like citrus fruits, tomatoes, fruit juices and drinks, colas, wine, pickles, vinegar, or drinks with caffeine or alcohol, the acids' irritating effects on the stomach are intensified.

For most adults, the doses listed on the label of the aspirin bottle are safe. But the PDR says that people who are allergic to aspirin or who have asthma, stomach problems that persist or recur, gastric ulcers, or bleeding problems should not take aspirin unless directed to do so by a doctor.

As for aspirin plus alcohol, the Food and Drug

Administration has recommended that anyone who regularly has three alcoholic drinks a day should check with a doctor before taking aspirin or any other over-the-counter painkiller. The FDA decided that alcohol plus aspirin was risky mostly because they both irritate the stomach lining in the same way. And a small 1990 study found that aspirin significantly lowered the body's ability to break down alcohol in the stomach, resulting in blood alcohol levels that were 30 percent higher than those in people who drank alcohol alone.

HIGH ON COFFEE

Q. Why is caffeine a stimulant?

A. Caffeine is not a direct stimulant; instead, it blocks the action of another chemical, naturally present in the human body, that has a calming effect on the activity of cells, especially those in the brain and spinal cord.

Caffeine, found in tea, coffee, and cocoa, is one of a class of chemicals called methylxanthines that temporarily occupy the same cell receptors as adenosine, a chemical produced by the body's energy metabolism. Adenosine is a neuromodulator, a substance that normally acts as a brake on nerve cell activity by keeping cells from staying excited too long. When adenosine's action is blocked by caffeine, nerve cells are more active.

Caffeine enters the brain very rapidly and stays active for several hours, but its effects drop off more quickly than those of some other stimulants. The time needed for caffeine concentrations in the bloodstream to drop by half is about three to six hours in a healthy adult, compared with about ten hours for methamphetamine.

Navy scientists have studied caffeine's short-term stimulation for possible use in military campaigns to counteract the effects of fatigue. The Operational Clinical Research Center of the Naval Aerospace Medical Research Laboratory reviewed the literature on the known effects of caffeine and concluded that "there is consistent and convincing scientific data that indicates the effectiveness of caffeine in counteracting the increased sleepiness and decreased alertness associated with sleep loss."

PRIMING THE HEAT PUMP

Q. Why does drinking alcohol give the illusion of warmth without keeping you warm in cold weather?
A. The assumptions of the question are not totally true. It's not just drinking, but where you drink and what kind of clothes you have on that determine whether the body ends up warmer or colder.

What happens is that alcohol causes the blood vessels in the skin to dilate, so that more blood passes through, and more blood means more heat going through the skin. So if you drink and are sitting in the living room

with a sweater on, for example, that heat due to vasodilation will be retained, and you will feel warm.

But if you go outside without out enough clothing, all that heat will dissipate. Such heat loss poses a serious risk of death from hypothermia for people drinking outside in cold weather. Excessive drinking also impairs sensation and the judgment that would ordinarily tell a person that it is time to come in from the cold.

CALCIUM WITHOUT COMPLAINT

Q. What form of calcium supplement is most easily digested? And why do the supplements sometimes cause indigestion?

A. Calcium citrate, the type found in calcium-enriched orange juice, is less likely to cause stomach upset than other forms. Calcium carbonate and calcium gluconate are the two forms that are most common in supplements; they both tend to be adequately absorbed. Calcium citrate is a good alternative, because it is very well absorbed and also enhances iron absorption, which may be inhibited by some other forms.

For all kinds of calcium supplements, it has been found that vitamin D, either in foods or supplements or manufactured by the body in the presence of sunlight, increases absorption of calcium.

The 1994 National Institutes of Health Consensus Development Conference statement on calcium says that gastrointestinal side effects of calcium supplements are usually observed at relatively high dosages. Calcium ions stimulate gastrin secretion and gastric acid secretion, which can produce a rebound hyperacidity when calcium carbonate is used as an antacid. These side effects should not be major problems with a modest increase in calcium intake, the statement said.

As with other nutrients, the conference emphasized:

"The preferred approach to attaining optimal calcium intake is through dietary sources."

Tea for Therapy

Q. I bought some milk thistle tea, *Silybum marianum*, at the health food store. It's supposed to be good for the liver. Is it helpful or harmful?

A. Milk thistle does contain chemicals called flavones that appear to benefit the liver, but these compounds are not very soluble in water, and the tea has only about one-tenth the strength of the original plant material.

In fact, the chemicals are poorly absorbed through the gastrointestinal tract and would probably be medically effective only if concentrated and injected.

Toxic effects have apparently not been reported from consumption of milk thistle. As with many herbal remedies, the concentration of active ingredients varies as much as fourfold from sample to sample, so dosage would be hard to control.

Milk thistle is a Mediterranean plant, a tall herb with prickly leaves and a milky sap. It has become naturalized in California and the eastern United States. It has small, hard fruits resembling seeds. While it was long used for liver complaints, it was virtually ignored in the first half of this century.

A few decades ago, however, German scientists isolated a crude mixture of chemicals from the fruits that appeared to have liver-protectant properties and called the extract silymarin. Silymarin was subsequently found to contain a large number of flavones, some of which protect against a variety of toxins, even the poison of the death cap mushroom. There have even been human trials for conditions like hepatitis and cirrhosis, with some encouraging results. Silymarin not only protects intact liver cells by blocking the entry of toxic substances, it also appears to stimulate protein synthesis and thus accelerate regeneration of liver cells.

COLONIZING THE GUT

Q. If I eat a lot of yogurt, will the bacteria in it make me sick?

A. There is no evidence of human health problems from yogurt with live bacteria, and there is some evidence the bacteria may be beneficial.

It is controversial whether yogurt bacteria survive in the gastrointestinal tract, and conclusions must await completion of research, but there is better-than-anecdotal evidence that bacteria can help colonize the tract and keep out bad bacteria. The theory is supported by studies of young chicks sprayed with bacteria present in adult chickens. The chicks take up these microbes and are not colonized by harmful salmonella, an effect called competitive exclusion.

There is no reason to eat only yogurt without live cultures for fear of danger from the bacteria that ferment milk into yogurt. The most commonly used culture is *Lactobacillus delbrueckii bulgaricus* in combination with *Streptococcus thermophilus*. Both are harmless in the gut, if they survive.

Q. If large fish, like tuna, can have traces of mercury, should I cut out tuna fish sandwiches?

A. There is both concern and controversy over mercury in fish, but an occasional tuna fish sandwich for an adult who is not going to become pregnant is not the main worry.

The biggest concern is the risk to a fetus. Mercury builds up in muscle cells of fish and in those who ingest it, and can remain to damage the nervous system of a fetus even months after ingestion, but what levels are dangerous has been debated. Recently, after an eighteen-month study, the National Academy of Sciences put the boundary of concern at levels only one-fifth as high as other proposed standards: 0.1 micrograms per kilogram of body weight a day. Translated to consumption of fish, to be on the safe side, a potentially pregnant woman might want to eat no more than one tuna fish sandwich a week. But the safe limit for other adults is probably much higher.

The riskiest fish are sharks and swordfish. Just behind them are the large species of tuna. But canned tuna uses much smaller species, with a smaller range of mercury accumulation, and so canned tuna is safer than tuna steaks.

Chalk Makes Its Mark

Q. What is kaolin, and where does it come from? Is the kaolin in diarrhea medication the same as the clay in china?

A. Kaolin, also called China clay, is a chalky rock composed chiefly of kaolinite, together with quartz and mica. It is formed by the weathering of aluminum-rich silicate rocks, especially feldspar. Over the ages, the crystals wash into large sedimentary deposits.

Purified kaolin in a suspension with pectin from apples has long been used to fight diarrhea. It is a gastrointestinal adsorbent, believed to attract and hold microbes that may be causing the diarrhea. The action of pectin is not known.

Kaolin is said to derive its name from an area of China with extensive deposits of the mineral. It is found in many other places famous for china, like Meissen, Germany; Limoges, France; and parts of England.

There are also kaolin deposits in the United States, including a great chalk mark across the state of Georgia. The China Clay Producers Association says the Georgia deposit dates from 50 million to 100 million years ago, when the Atlantic Ocean covered much of Georgia south of a line drawn from Columbus to Augusta. Above this line, known as the Fall Line, crystalline rocks from the Piedmont Plateau began to break down, and streams carried the tiny crystals seaward. The deposits were later covered with heavy layers of earth.

SEEKING A SIESTA

Q. Why am I so sleepy and drowsy after lunch? It makes it hard to concentrate.

A. There are three likely explanations, based on recent research.

First, what do you eat and drink? A lunch high in car-

bohydrates, especially if no protein is eaten along with bread and milk, is likely to produce calm to the point of sleepiness, because of the complex role of carbohydrates in shifting the balance of neurotransmitters in the brain. Conversely, some protein contributes to alertness. A three-martini lunch also induces sleepiness.

Second, as siesta-loving societies have long recognized, sleep researchers have found that there is an inborn readiness in human beings to take a midafternoon nap. Volunteers isolated from all evidence of day and night and left to set their own sleep patterns tended to settle into one long sleep session and a session of 1 or 2 hours about 12 hours after the middle of the main period.

Third, you may be sleep deprived. While the average adult needs 8 to 9 hours of sleep a night, most get only 7, and many get 6 or less during the workweek. People permitted to sleep as much as they choose sleep 10.3 hours out of every 24, just as monkeys and apes do.

An afternoon nap may be just what nature ordered to restore alertness.

CHAPTER 14

Battleground: The Body

WHERE DANGER LURKS

Q. How long can germs live on an object? Does the surface have to be moist? Sometimes when I get recurring colds or flu, I wonder if I am reinfecting myself.

A. Some disease-causing organisms, or pathogens, can thrive and multiply for some time on an object like a phone receiver spattered by the saliva of an infected person. But most bacteria and viruses quickly die when moisture disappears, usually after one or two hours for the saliva spray.

Different pathogens have different survival times, ranging from a few minutes or hours to days or even months, and some might lurk for weeks or months on an object like a door handle in the house of someone with hepatitis A. In such cases, infectious disease specialists recommend frequent hand washing and avoiding touching your eyes or mouth; such measures would cut your infection risk to a very low level.

Using the damp toothbrush of someone with some infectious diseases would be a very likely way to get sick yourself, but cold viruses are usually spread by hand-to-eye or hand-to-mouth contact.

A Germ's Germ

Q. Do germs have germs?

A. There are viruses that infect disease-causing bacteria, as well as many other bacteria. Called bacteriophages, meaning bacteria eaters, they were discovered during World War I and named in 1917.

Much smaller than the bacteria they attack, bacteriophages cannot grow and multiply on their own. Instead, like other viruses, they attach themselves to the surface of the target and insert their own genetic material, taking over the cellular machinery and using the bacteria's

chemical energy to build and assemble the parts of new bacteriophages.

The resulting material soon reaches a volume that weakens, then bursts, the walls of target bacteria; this process is called lysis.

Bacteriophages, which are very common, are highly specific in their targets, with each kind attacking only certain species of bacteria.

Bacteriophages have been investigated as possible alternatives to antibiotics. The ones that attack the useful microbes *Lactobacillus acidophilus* and *Lactobacillus caseii,* used as starter cultures in fermented milk products, have also been studied by researchers who want to foil the attack.

PINK AND PURPLE BACTERIA

Q. What do doctors mean when they talk about gram-negative bacteria?

A. They are referring to how the germs react to one of the basic tests in bacteriology, the Gram stain or Gram's stain, which distinguishes between two major classes of bacteria by how they take up certain dyes.

On a microscope slide, heat-treated gram-positive bacteria take up a purple dye (originally gentian violet), which is then fixed with an iodine solution, and they hold on to the color even after the slide is washed with a solvent. Gram-negative bacteria, which have thinner but more complex and less permeable cell walls, do not retain the purple dye, but take up a counterstain, safranine, and end up a dark pink.

Gram-negative bacteria are of concern because they tend to be more resistant to antibiotics. Many cause serious diseases, including the plague, Legionnaire's disease, and many bowel-related diseases, including cholera and salmonella.

Opportunistic infections from gram-negative bacteria are also a major danger in hospitals. When some of them escape from the gut and enter a patient's bloodstream through medical procedures, they can cause septic shock, which is often fatal.

The test was named for Dr. Hans Christian Joachim

Gram, a Danish doctor who developed it while working at the University of Berlin in 1884. He was trying to distinguish between two kinds of organisms that cause pneumonia: pneumococci and *Klebsiella pneumoniae.*

MASKING A COLD

Q. When Asian people wear face masks, is it to keep from getting a cold or other infection, or is it to keep from spreading one? Does it work?

A. Polite East Asian people with a cold may wear a mask with the aim of preventing its spread; this is especially common in Japan. In China, the idea may be to cut down infection or it may be a reaction to high levels of visible pollution; in some of China's industrial areas, masks are worn simply in the hope of keeping coal dust out of the respiratory tract.

However, there is little hope of cutting down on the transmission of colds by wearing a mask. Medical authorities believe that most colds spread from nose to hand and then from hand to nose, rather than through the air, so a considerate cold victim should instead wash his or her hands frequently.

The influenza virus does travel easily through the air as a passenger on water droplets, so a mask may head off its transmission.

WASH YOUR TROUBLES AWAY

Q. Is there any way to keep from getting a cold?

A. There are no guarantees, but there are some ways to cut your risks. The most important are avoiding crowds, especially during the peak cold seasons of September, late January, and April; avoiding people who obviously have colds; avoiding shaking hands with people who might have them; avoiding rubbing the nose and eyes; and washing your hands, washing your hands, washing your hands.

These precautions may help because of the way colds are often spread: not just through infected droplets in the air but through particles carried by hand to eye and nose membranes.

The greatest share of colds are caused by rhinoviruses (about a third) and coronaviruses (about 10 percent). Rhinoviruses seem to spread best in droplets from a runny nose, though they may also be sprayed by coughs and sneezes.

Once virus particles land on a surface or someone's hands, they may remain active for several hours. Then the virus can go from hand to face to eye and nose membranes.

Someone with a cold can be contagious even before signs of illness appear and for a couple of days after it is obvious, so one cannot entirely prevent colds by avoiding people with symptoms, but it helps. Other useful measures are eating a balanced diet, which is associated with a lower suscepti-bility, and staying away from children, who get five to eight cold infections a year. Those younger than 2 have the most.

Q. Can you reinfect yourself with an infectious disease from which you are recovering, perhaps through exposure to residual germs on your hands?

A. There is no simple answer, but in general, this is very unlikely. In theory such reinfection is conceivable, but there are no documented cases of it.

The possibility would depend on the organism and how the person's immune system responded to it. If a pathogen is on the skin a long time, the person has probably developed immunity to it. The same holds true for an infection from which a person is recovering, so it is difficult to "reinfect" yourself unless it is a totally different organism.

Even if a person has two cases of flu in the same winter, the second case would probably involve an unrelated flu virus that was present at the same time. The same patterns of disease would appear, so it would be impossible for the patient to distinguish between the two infections.

Long-Lasting Immunity

Q. Why does immunization have long-lasting effects?

A. Immunity is long-lasting because some of the many kinds of cells that make up the normal human immune system are naturally able to retain a "memory" of the defense efforts aroused by certain infections or vaccines.

There are many kinds of immune cells, including T cells and different types of white cells. Their interaction with disease is complex and involves long chains of interactions and responses.

A vaccine fools the body into thinking it is being infected and creating a defense against that organism that it keeps in its immune memory. Later, if exposed to the infection, the body fights it quickly.

For example, after an actual infection with a disease like measles, which people rarely get more than once, immune memory is created, and certain cells are ready to fight again upon reexposure. The memory remains for years.

For some vaccines, protection may last a long time, even a lifetime. But for others, vaccines may protect for only a short time.

PUTTING VIRUSES TO WORK

Q. Unlike bacteria, which can be useful in many ways, there don't seem to be any "good" viruses. Do any play a positive role?

A. From the human point of view, some viruses can be useful, potentially useful, or at least interesting.

Under the "interesting" heading would fall the mosaic virus, which produced the fantastically variegated tulips of the tulipomania investing craze in seventeenth-century Holland. The rare infected bulbs were treated almost like a currency with a constantly multiplying value until the market crashed in 1637. It was discovered that the infection weakened the plants, and now hybridizers work to create the same colorful effects by breeding rather than infection.

Unquestionably useful in the field of public health is the vaccinia virus. It was discovered that milkmaids infected with relatively mild cowpox were immune to virulent smallpox, and the knowledge eventually led to widespread vaccination, eradicating smallpox as a disease. Later, the related vaccinia virus was used for this purpose.

In genetic medicine experiments, harmless viruses are used to carry desirable genes into cells. For example, scientists have tried treating a genetic disease called familial hypercholesterolemia by letting a virus insert a crucial gene into liver cells that makes them produce a chemical sponge for harmful cholesterol.

THE SICKNESS SEASON

Q. Why is flu season in winter? Why in general are more people ill in winter?

A. It is not cold feet and wet heads that are the problem, infectious disease experts say, but the fact that human beings are warmth-loving social animals. At least in cold climates, widespread outbreaks of diseases like colds and influenza tend to start in winter months, when people spend more time together indoors in close quarters with the windows shut. The cold months also bring children, those well-known vectors of bacteria and viruses, together in the classroom and day-care center, where they can pick up infections and take them home to the rest of the family.

In the Northern Hemisphere, early winter also includes major holidays with their attendant crowds and socializing. And even air-kissing under the mistletoe and

New Year's handshakes may help transfer germs, some of which travel in airborne respiratory particles while some are spread from hand to hand.

ALTITUDE ADJUSTMENT

Q. I am physically fit, but I almost always suffer altitude sickness on skiing weekends. What can I do?

A. Try a longer vacation. Altitude sickness is not linked to a person's general condition but to specific adaptations to high altitudes; even puny permanent residents are usually immune.

Medical experts say the body usually adjusts within a few days to the complex changes in blood chemistry that come about because of the relative shortage of oxygen on high slopes. It produces more oxygen-carrying red blood corpuscles and more capillaries to carry them. The experts also recommend a gradual ascent, if possible. Some recommend medications, including diuretics, to ease the adjustments.

If you get symptoms of what is also called acute mountain sickness (including some that are often mistaken for flu), you should seek oxygen, move to a lower elevation, avoid exertion, and drink plenty of fluids. In 1992, researchers working with astronomers on 13,796-foot Mauna Kea in Hawaii found that floods of water helped ease splitting headaches in many healthy workers. Dehydration occurs because the body tries to thicken its blood to improve absorption of the scanty oxygen.

The Mauna Kea researchers also recommended a breathing technique in case of faintness, disorientation, or nausea: Take a deep breath and hold the nose and mouth tightly closed while pushing to expel the air from the lungs. The increased pressure of the air in the lungs is enough to drive more oxygen into the bloodstream.

FELINE FEVER

Q. What is cat scratch fever?

A. Cat scratch fever or cat scratch disease is a mild flu-like infection, usually causing swollen lymph nodes and a low-grade fever, that was first described in the 1950s as being associated with cat scratches, especially from kittens.

The infection usually goes away by itself in a few weeks and can be treated with antibiotics, but it can cause a severe inflammation called bacillary angiomatosis in patients with weakened immune systems.

A cat carrying the microbe does not show symptoms, and medical and veterinary authorities say it is not necessary to get rid of it. If someone in the household is at high risk, a test to detect the infection can be done and the cat can be treated. People who have not been exposed to cats can also be infected.

It was difficult to find the disease organism, and it was long assumed to be a virus. A small bacterium called *Afipia felis* was mistakenly identified as the cause in the 1980s, but later research found the disease was caused by a bacterium called *Rochalimaea henselae,* eventually reclassified as *Bartonella henselae.* It was named for Dr. Diane Hensel, a microbiologist who did some of the research.

The bacterium is closely related to *Bartonella quintana,* which caused trench fever in World War I and is spread from person to person by the human body louse. In cat scratch fever, however, it appears that fleas spread the infection from cat to cat, but not from person to person.

RESPIRATORY DISTRESS

Q. I know ARDS is a respiratory disorder, but what type?

A. ARDS, which stands for acute respiratory distress syndrome, or sometimes adult respiratory distress syndrome (to distinguish it from a lung problem in newborns), is not a disease itself, but a type of severe acute lung dysfunction that can result from disease or injury, according to the ARDS Support Center, a clearinghouse for information.

The condition stiffens the lungs, fills them with water, and causes shortness of breath from respiratory failure. Inflammation damages microscopic air sacs called alveoli, which collapse, and tiny blood vessels called capillaries, which leak fluid into the lung. The inflammation can cause permanent scarring.

When doctors first became aware of ARDS in World War II, it was invariably fatal, but modern mechanical respirators, delivering extra oxygen, now help many to survive it.

Treatment of the condition can cause its own damage, like oxygen burns, and may leave long-term problems.

Doctors have learned to look for ARDS and treat it in patients who have survived auto accidents or heart surgery or who have illnesses like pneumonia or severe infections like septic shock, but they do not know why some of these patients suffer ARDS and others do not.

MAD COWS AND YOU

Q. Can foot-and-mouth disease and mad cow disease be spread to humans by dairy products?

A. Foot-and-mouth disease is a danger only to animals, not to humans, and the chance of getting mad cow disease from dairy products is zero to almost zero. The risk is so slight that there is an exemption allowing the importation of milk from countries where mad cow disease has been found.

Transmissible spongiform encephalopathy, or TSE, is the umbrella term for illnesses—like bovine spongiform encephalopathy (the formal name of mad cow disease) and Creutzfeldt-Jakob disease—that are believed to be caused by prions. Neither viruses nor bacteria, prions are misfolded proteins. Other such diseases are scrapie in sheep and chronic wasting disease in deer and elk.

The prevailing theory is that a prion acts like a seed crystal, causing healthy proteins, usually in the brain, to misfold and bring on destruction of brain tissue, creating spongelike cavities. But the only tissues in diseased cattle likely to contain dangerous levels of prions are those in and around the nervous system and the very end of the small intestine.

AIDS AND MOSQUITOES

Q. What evidence do scientists have that mosquitoes don't carry HIV from person to person?

A. There are three kinds of evidence. First, there is epidemiological data from studies of AIDS outbreaks. For example, in sub-Saharan Africa, where mosquitoes are extremely plentiful, the very people who would get bitten most, young children who are outside playing, comprise the one group that is not HIV-infected. The AIDS epidemic is among the sexually active and newborns, except cases involving transfusion.

The mosquito is also incapable of biological transmission of the virus. HIV does not grow in any mosquito cells, and the life cycle component of biological transmission is not there, in contrast to malaria, for example, which depends on the mosquito. HIV cannot grow outside cells, and if the viruses don't infect any mosquito cells, they can't use the mosquito as a biological host.

What about mechanical transmission, in which a trace of HIV-infected blood could be transferred to the next person bitten? Epidemiologists consider what is known about the amount of blood necessary to transmit HIV through a needle stick, which is a rather large amount, about a tenth of a milliliter. Given the size of a mosquito's stinger and the amount that could reasonably be carried from one bite to another, thousands of mosquitoes all biting at once would be required to deliver the amount of blood necessary to pass on the virus.

AIDS IN AFRICA

Q. Why is AIDS so prevalent in Africa?

A. There are many theories about the high AIDS rates, especially in sub-Saharan Africa, and answers almost certainly lie in combined economic, sociological, political, and medical factors. For example, the rate of HIV infection in Botswana, the highest in the world at 35.8 percent of adults, has been linked to Botswana's location and excellent road system, making it a crossroads for sexually active truck drivers.

For another example, drugs to prevent transmission by mothers to infants are too expensive for most, while testing costs mean people with no symptoms spread HIV, unaware they are infected.

The highly different rates of AIDS from region to region within Africa have made it a laboratory for studying control policies. Factors that may theoretically explain the rate differences include different patterns of sexual activity; the frequency of other sexually transmitted diseases, especially those that break the skin; the presence of other infections such as malaria or tuberculosis that can increase the amount of HIV in the blood; differences in circumcision frequency; virological differences in the strains of AIDS; other behavior differences; and policy differences, like reducing the stigma of AIDS and promoting education on safe sex.

Rodents and Rabies

Q. If the rabies virus can be transmitted by any mammal, why do we not hear of its being spread by rats in the United States?

A. Scientists don't know for sure, but the most likely explanation is that rats probably would not survive an encounter with an infected carnivore, and so would not be around to transmit the disease to people. It is also possible that rats may lack a means of transmitting the virus within the species, because of mouth shape or some other factor, and so do not become significant reservoirs for the disease. There is not much fighting within the species that would lead to the bites that would spread the infection.

The most common wild reservoirs of rabies are raccoons, skunks, bats, foxes, and coyotes. Domestic mammals, including cats, cattle, and dogs, can also get rabies. But small rodents such as squirrels, mice, rats, hamsters, guinea pigs, gerbils, and chipmunks and lagomorphs such as rabbits are almost never found to be infected with rabies and have  not been known to cause rabies among humans in the United States. From 1985 to 1994, woodchucks (or groundhogs) accounted for 86 percent of the 368 reported cases of rabies among rodents.

Q. My father told me to hold my breath when a bus passes. Would this do any good in terms of protecting my lungs from pollution?

A. Probably not, but more research is needed to know the degree of danger presented by the fumes emitted from diesel vehicles.

Cars, trucks, and buses, especially diesel buses, emit large amounts of tiny particles that remain airborne for long periods. Because they linger, holding the breath until the bus passes is ineffective in keeping them out of the nose.

Small particles of less than 2 microns (two one-thousandths of a millimeter, or roughly 0.00008 of an inch) predominate in vehicle exhaust, and because of their small size, they are not filtered out by the nose but are able to pass easily and deeply into the lungs. They are mostly carbon soot, but have a wide range of volatile organic compounds stuck to them. Among them are the polycyclic aromatic hydrocarbons, PAHs, which are known or suspected carcinogens. The particles also contain irritant compounds that can acutely affect the lungs.

ALLERGIC TO WINTER

Q. My friend insists she is allergic to winter. Is that possible?

A. Conditions that are very like an allergy to winter are not only possible but common.

Your friend's problem could be an asthmatic reaction to exercising in the cold, or it could be an allergy to something she is exposed to in winter even more than in the traditional allergy season. It could even be that a humidifier she is using to fight winter dryness is spewing out allergens, the proteins that cause the body to react with the symptoms of an allergy.

Exercise-induced asthma symptoms are usually worse in cold weather and during activities that stir up a cold wind. Although the precise cause is not known, it is believed that they are set off by the drying and cooling effects of air moving rapidly in and out of the respiratory tract.

Allergies that are active year-round are called perennial allergic rhinitis. The symptoms can mimic those of the common cold. The American Academy of Allergy, Asthma and Immunology says that among the most common culprits are dust mites, animals, and mold spores.

Mold spores, the highly irritating feces of dust mites, and the dander of cats and dogs, trapped in a tightly sealed house, can be stirred up when the furnace goes on. And some humidifiers can breed molds themselves or add so much humidity that molds thrive elsewhere in the house.

CHAPTER 15

Ways Modern and Ancient

Hot Stuff

Q. Why do some kinds of cookware get very hot in the microwave?

A. There are two reasons why cookware may heat up excessively in a microwave oven. The first is the same effect that occurs in cooking: friction. Microwaves are rapidly fluctuating electromagnetic fields, which push on electric charges and twist naturally polarized particles like water molecules back and forth. In liquid water, this twisting leads to molecular friction and heating. If the cookware is wet, like a porous crockery cooker that is moistened before use, it will get very hot.

A second effect can also cause trouble as the microwaves push free electric charges around. In a good conductor of electricity through which the charges move easily, the microwaves are reflected. The metal walls of the oven, for example, barely heat at all. But in a poor conductor, the charges being pushed around in the material cause resistance heating, just like that in the heating element of a hair dryer.

To test pottery or dinnerware for use in a microwave, put the empty dish in the oven alongside a cup of water in a microwave-safe glass container. Microwave on high for one minute. If the dish stays cool, it is safe to use.

Don't use porcelain cups with gold rings around the lip. If you try to reheat your coffee in one, the ring burns up just like a lightbulb filament.

SEEING STEAM

Q. Why do I see more steam when I take the teakettle off the flame?

A. The "steam" you see is what steam engine enthusiasts call wet steam, because it contains unevaporated water as a mist within the actual water vapor, which is a colorless, odorless gas. Dry steam does not contain any unevaporated water.

Steam vapor forms at temperatures above the boiling point, normally 212 degrees Fahrenheit. Visible steam is either water that is not yet hot enough to form this vapor or, in the case of your kettle, water that has cooled slightly into water droplets.

When a whistling teakettle is taken off the heat, the

whistle gets weaker, because the pressure of the water vapor drops as the heat drops, although the visible water droplets momentarily increase.

FIGHTING FOG

Q. How do antifogging treatments for mirrors and eyeglasses work?

A. Antifogging agents are chemicals that force the tiny droplets of water that make up fog to merge into a transparent sheet. An agent is sprayed as liquid on the lens, and when it dries, it forms a clear film. Normally, water beads up, but with the film, the beads run together. The water is still there, but you can see through it.

Such agents are surfactants—chemicals that lower the surface tension of the liquid with which they are in contact, in this case water. Soap and detergents contain surfactants.

A low-cost emergency substitute for commercial antifogging agents is potato juice. In research on land mine removal at the University of Western Australia, potato juice was tested for use in poor countries to keep blast-protection visors from fogging up. It performed as well as some brand-name products.

The researchers cut a slice from a raw potato, rubbed it generously and evenly onto the inside surface of the visor, then wiped away solids and excess juice with a soft cloth.

OLD NEWS

Q. Why does paper, especially newsprint, turn yellow as it ages?

A. The paper is oxidizing, one of several destructive processes that affect paper. Newsprint is high in lignin, one of the main structural components of wood. When lignin is in contact with oxygen, a highly reactive chemical, the reaction changes the way the fibers reflect light, discoloring the paper. Oxidation also produces more acid, a major culprit in the breakdown of many modern papers.

Older paper was usually made from linen or cotton fibers, which required little treatment to offer a good writing surface. Wood is much cheaper, but manufacturers found they had to use a thin chemical layer called sizing on the paper surface to make it less absorbent and prevent blurring when ink is applied.

A commonly used sizing is a compound of alum, or aluminum sulfate, and when exposed to warmth and high humidity, alum molecules split up and form an acidic solution. Cellulose fibers, which are made of molecular chains of carbon atoms, are easily split by even weak acids. The strength of the fibers is destroyed, and the paper becomes fragile.

The acid problem affects nearly half the books published since 1850, conservators estimate, and it was not until the 1950s that publishers began to abandon alum sizing for alternatives to make some paper acid-free. Conservators are also working on treatments to deacidify paper.

THE BRAIN AS 3-D CAMERA

Q. How do those 3-D pictures work, the ones for which you focus your eyes to see hidden images?

A. They work somewhat the way an old-fashioned stereopticon viewer (holding two slightly different photos) works. Each eye focuses on a separate image, but the brain combines them into one image that gives the illusion of depth (the third dimension) in addition to width and height. It does this by matching up the similarities and adding in the small differences. The result looks like what you see when you look at a solid object from the two slightly different angles of your two eyes.

The first pairs of random dot stereograms, with images embedded in dots, that can be seen only by people with good depth perception, were produced in

1959 by Dr. Bela Julesz; twenty years later, Christopher Tyler, his student, found that patterns could be arranged in a single image to get the same effect.

The ability to see computer-generated 3-D images depends on having healthy vision with both eyes working smoothly together, and the images can be used as a test or exercise. Vision problems that can interfere with depth perception include amblyopia, or lazy eye, and strabismus, informally called walleyes or crossed eyes.

THE RIGHT TOUCH

Q. When using an ATM, sometimes you don't have to touch the screen, just place an object close to it. How does that work?

A. Several kinds of technology have been used for touch screens. If the screen responds without actually being touched, it probably uses a system that has tiny infrared light emitters and detectors along the edge of the screen. When a finger interrupts the grid of beams, a computer can tell its exact position.

Some screens sensitive to actual touch use low-voltage electrical fields; a touch completes a circuit, giving a readable location.

EXPLODING ELEVATORS

Q. Why is grain dust in elevators explosive?
A. Just about any organic material will explode if it is fine enough, dry enough, in suspension in air, and confined. All it takes is the fuel, oxygen, and a spark.

Sugar, coal dust, wood dust, walnut hulls, the ingredients of children's modeling dough, and some medications have also been involved in recent dust explosions. Organic particles vary in flammability, and wheat starch and cornstarch are more explosive than other kinds.

Often a primary explosion shakes the building, dislodging more dust, which ignites in turn. There is also a pressure wave ahead of the flame front that shakes loose still more dust. The stronger the  elevator, the greater the danger—a stronger container lets pressure build up so that when it is finally released, the force is much stronger.

Proper venting to relieve pressure is one important safety measure. Others are cleaning grain of foreign material and broken kernels before storage, spraying edible mineral oil on conveying equipment to control dust, and installing pneumatic dust-collection systems using bag filters.

LIGHTNING TARGET

Q. Do buildings in Manhattan need lightning rods? I never see any.

A. Historically speaking, New York City is less prone to thunderstorms than many areas of the country, but the island of Manhattan is not immune to lightning. For example, the Empire State Building is struck by lightning on average about two dozen times a year.

The good news for Manhattanites is that modern high-rise buildings have built-in lightning protection systems. Because the highest object in the area runs the greatest risk of being struck, the tall, protected buildings help shield lower structures.

Lightning is an electric charge trying to complete a circuit, and it seeks the shortest route to earth to do so. A lightning protection system provides it with such a path to the ground, carrying the current around the structure.

The best time to plan the system is when the building is still on the drawing board. A complete lightning protection system, consisting of air terminals, bases, conductors, and groundings, can be concealed within the building's framework.

People who are outside in a thunderstorm, even in

Manhattan, should take shelter in a substantial structure (a bus shelter won't do) or failing that, a fully enclosed metal vehicle. Inside, people should stay away from water, windows, phone lines, and appliances until 30 minutes after the last thunderclap.

SWAYING IN THE BREEZE

Q. The last time I went to the Empire State Building, it looked as if the building was swaying. Is this an illusion?
A. The building does move, but the motion is a very slight bending, only about a quarter of an inch, in winds of more than 75 miles an hour. The illusion of a larger motion probably has to do with wind-whipped clouds; an observer on the ground can see a tall building appear to "fall" if the clouds above it are being blown in the right direction.

Older skyscrapers like the Empire State Building and the Chrysler Building would sway some inches or even feet if they were not stiffened with a massive sheathing of stone and masonry, engineers say.

Newer skyscrapers, with their barely covered skeletons of steel, may sway perceptibly. The 579-foot Trump International Tower, the old Gulf and Western Building on Columbus Circle, was stiffened to defeat a perceptible twist in high winds of 2½ feet; a 44-by-15-foot brace from the street to the roof was added. The reason is comfort, not safety, the developers say, because the amount of sway that an office worker can ignore would be disturbing to a full-time residential tenant.

Structural engineers, who are not concerned about the movement's impact on the building's own integrity, use a number of techniques to dampen the motion so that people inside a tower do not perceive it. Current strategies include making buildings heavier by using more concrete, cross-bracing the steel framework, or devising a mechanical system that uses rooftop weights to offset motion.

BURNING ISSUE

Q. Is spontaneous human combustion real?

A. There is no scientific evidence that any person ever spontaneously combusted, and no scientifically known means by which it could happen.

In a typical case where there has been a report of a body burning unusually, a simple or at least plausible explanation is available when the case is examined in detail.

A statistically typical case might involve an elderly or infirm woman who is alone and near a fire source at night and in winter. The fire typically starts on or near the person, and it often involves the clothing, chair, or bed. In all cases there are fuel sources, a plausible explanation for the ignition, and sufficient oxygen.

The idea that the body could spontaneously ignite was promoted by the nineteenth-century temperance movement. Leaders of that movement correctly noted that many fatal burning cases involved drinkers.

THE GREEKS HAD A NAME FOR IT

Q. Was there a secret Byzantine naval weapon called Greek fire?

A. Historians agree Greek fire existed, but debate its nature. Some credit Callinicus of Heliopolis with building fire-spouting ships that defended Constantinople from the Arabs in 673. Greek fire was described as a liquid that burned even on the sea surface, with smoke and whooshing sounds.

Candidates for ingredients include petroleum, resin, and quicklime, plus an oxidizing ingredient, possibly saltpeter, which would make Greek fire an important precursor of gunpowder.

One older theory was that rather than using saltpeter or some other oxidizer, the Byzantines instead distilled petroleum to make kerosene or gasoline. But more recent research suggests that undistilled petroleum was used, collected in the cooler months so that its volatile fractions were less likely to evaporate, and kept in heated containers on board ship.

LETTING IT ALL OUT

Q. Did bleeding the patient actually help any medical condition?

A. Not unless the victim—that is, patient—chanced to be suffering from a hypertension crisis or an abnormal buildup of iron in addition to whatever disease the doctor thought he was treating.

For example, draining 10 to 12 ounces of blood, for a start, and up to 80 ounces in all, was routinely prescribed during the American Revolution for yellow fever among Washington's troops. The practice, called phlebotomy, is unlikely to have done the soldiers any good, especially if they were already weakened by cold, malnutrition, purges, and enemas.

Bleeding or bloodletting was common for centuries, if not millennia. It was based on the ancient theory that bodily humors, called black bile, yellow bile (or choler), phlegm, and blood, had to be kept in balance. (The humors were also thought to determine temperaments, causing people to be choleric or sanguine, for example.) If a person was ill, the theory went, balance had to be restored by reducing whatever humor was thought to create the plethora, often by draining off blood.

Finally, in the nineteenth century, a French pathologist, P.C.A. Louis, decided to find out if bleeding really helped, by using what was then a revolutionary method: counting who got better and who got worse. His "numerical method" was applied to 78 cases of pneumonia, 33 cases of erysipelas (a flesh-eating strep infection of the skin), and 23 cases of inflammation of the throat. He found no advantage to bloodletting, and it was the beginning of the end for bleeding and leeching as well. However, using leeches to reduce a bloody swelling is still a practical part of modern medicine.

THE AWKWARD AMPHORA

Q. Why did the ancient world use amphorae? They appear awkwardly shaped for stacking.

A. The long pointed bottom of the typical amphora actually gave it an excellent shape for stacking against a ship's hull, ensuring that the oil or wine within would be safe during transport. In the evolution of amphorae, they often began rounded and egg-shaped and shifted to pointed, because the long bottom of the brittle pottery jar better withstood the pressure of being pushed sideways when stacked.

In many shipwrecks, amphorae are found still neatly stacked, though the packing between them has usually disappeared.

Many amphorae have what are called ear handles: small, round handles about as big as the hole made by the thumb and forefinger. The shape doesn't make sense for carrying; some archaeologists think the amphorae were probably tied down in rows through the handles. No ropes have been found, however.

ANCIENT CENSUS

Q. How do archaeologists estimate past populations?
A. There are some rules of thumb for estimates for a limited area, but estimates for whole countries or the world are just guesstimates. Archaeologists tend to find and count all the formerly inhabited sites in a study area; make an educated guess about which ones were occupied at a certain period, usually based on pottery of a certain age found on the surface; and then from the total occupied area make a rough estimate. Evidence from aerial photos and satellite images may also be used.

From studies of modern villages, archaeologists can

say about 100 or 200 people occupy a hectare, or two and a half acres, of a modern settlement, and apply this estimate to the past. But there are tremendous uncertainties in such estimates. For example, big sites covered with the pottery of one period may indicate big cities, and some, like old Damascus and old Jerusalem, show evidence of densities of 500 to 600 people per hectare.

DIGITAL COMPUTING

Q. The number system with a base of 10 is so common that I wonder why so many societies developed it independently. And where did some of the other systems come from, like the one based on 60?

A. Historians of science find the obvious explanation for the base-10 system at the fingertips of human beings, all ten of them. Not so coincidentally, the word *digit* comes from *digitus*, the Latin word for finger or toe.

Counting on the fingers (base 10) or the fingers and toes (base 20) is the apparent root of many ancient written number systems. The French language retains traces of a base-20 system, based on multiples of 20, like *quatre-vingt*, or four 20s, for 80. The Maya also developed a base-20 system 1,500 years ago.

The Babylonian-Sumerian base-60 system survives in the measurement of time, angles, and geographical position; some scientists suspect this was also derived from finger counting. But others think it might have arisen because of trade with some other people with a system of weights based on a different, larger unit, resulting in the need to find a common way to express fractional weights.

PINPOINT QUESTION

Q. How many angels can dance on the head of a pin?

A. Medieval theologian-philosophers tried to calculate that incalculable number based on the notion that angels were the smallest possible physical creatures, though with very large spiritual powers. Based on that same definition of size, a modern physicist actually made a calculation for an even smaller dance floor: the number of hypothetical angels that could dance on the point of a pin.

The calculation was offered in 1995 by Dr. Phil Schewe, spokesman for the American Institute of Physics. He presented his idea at a meeting of the Society for Literature and Science.

For the smallest possible angelic size, Dr. Schewe relied on an idea drawn from superstring physics that space itself is not infinitely divisible, but breaks down at a distance scale of 10^{-35} meters.

For the size of a pinpoint, he took the tip of the IBM scanning tunneling microscope, the one that arranged 35 xenon atoms in the shape of the letters IBM. The tip tapers down to a single atom.

From there, it is an easy calculation, Dr. Schewe concluded. Figuring that the point is an angstrom across, or 10^{-10} meters, you divide that by 10^{-35} and the answer is 10^{25} angels (that's a 1 followed by 25 zeroes) can fit on the point of a pin.

Q. Can you be cursed for going into an ancient tomb?

A. You can certainly be cursed at, by archaeologists. The history of ancient Egypt, for example, is full of holes left by grave robbers, and remaining tombs are at risk from tourism and development.

Egyptians believed tomb inscriptions could confer

either blessings or curses on visitors, and early visitors probably believed so, too. Tomb builders wanted to encourage passersby to leave food for the dead or just to recite a standard ritual wish for bread, beer, cattle, and fowl, which, merely through the spoken word, would magically generate those commodities for the soul of the dead.

By 2000 B.C., standard inscriptions included both praise and blessings for those who recited the ritual and warnings cursing those who might damage the tomb or its images. The warnings invoked condemnation and even outright slaughter at the hands of the gods of the underworld. This was not an idle threat—there was a clear understanding that the dead could intervene in the lives of people on earth.

Notes on Sources

Chapter 1. *Reaching for the Stars*

Birth and Death: Nolan Walborn, an astronomer at the Space Telescope Science Institute.

In a Spin: Vera C. Rubin, an astrophysicist in the Department of Terrestrial Magnetism at the Carnegie Institution.

Death by Black Hole: Mitchell Begelman, professor of astrophysics and chairman of the Department of Astrophysical, Planetary, and Atmospheric Sciences at the University of Colorado at Boulder.

Averting Your Eyes: Alan J. Friedman, director of the New York Hall of Science in Queens.

Seeing Stars: Scott Kardel of the Lake Afton Public Observatory in Wichita, Kansas; Douglas Blanchard, chief of the Earth Science and Solar System Exploration Division at NASA-Johnson Space Center.

A South Star?: The Facts on File Dictionary of Astronomy; The International Encyclopedia of Astronomy (Orion Books).

Hello Out There!: The SETI Institute.

Sitcoms in Space: Seth Shostak, an astronomer at the SETI Institute.

Shine On, Harvest Moon: Alan J. Friedman; Joe Rao, lecturer at the American Museum of Natural History.

Mars Coordinates: Robert G. Strom, professor of planetary science at the University of Arizona.

Pale in Comparison: Joe Rao.

Cosmic Collision: Bryan G. Marsden, associate director for planetary sciences at the Harvard-Smithsonian Center for Astrophysics.

The Planetary Plane: Joe Rao.

Ring Around the Sun: NASA.

Space Traffic Control: Jonathan McDowell of the Harvard-Smithsonian Center for Astrophysics.

Steering the Shuttle: Kyle Herring, NASA Johnson Space Center.

Sci-Fidelity: John Lawrence, chief of operations and program support for the shuttle at NASA Johnson Space Center.

Getting Home from Mars: Mark Adler, Mars exploration program architect at NASA Jet Propulsion Laboratory.

Chapter 2. Down to Earth

From North to South: Phillip F. Schewe of the American Institute of Physics.

Rocks That Tell a Story: Understanding Earth, by Frank Press and Raymond Siever (W. H. Freeman).

Shocking Events: U.S. Geological Survey.

How Murky Water Gets That Way: Gulf of Maine Aquarium.

Putting the Pressure on Water: Alan J. Friedman, director of the New York Hall of Science in Queens.

Bends in the River: Understanding Earth, by Frank Press and Raymond Siever.

When the Equinox Comes Early: Joe Rao, lecturer at the American Museum of Natural History.

Chapter 3. The Weather, Rain or Shine

Laying Out the Odds: Sam Perugini, a meteorologist at Pennsylvania State University.

The Color of Clouds: Todd J. Miner, a meteorologist at Pennsylvania State University.

Cloud Capacity: Sam Perugini.

Trail of the Jet: Dr. Owen Brian Toon, professor of atmospheric and oceanic sciences at the University of Colorado.

Sowing the Skies: Weather Modification Association.

Measuring Rain: Todd J. Miner.

It's Raining Hamburger Buns: Todd J. Miner.

Bad Timing: Todd J. Miner.

Hot Air, Cold Peaks: James C. G. Walker, professor of geology at the University of Michigan.

Bending Sound: John Earshen, Angevine Acoustical Consultants of East Aurora, New York.

Invisible Ice: Fred Gadomski, a meteorologist at Pennsylvania State University.

Chapter 4. Food, Glorious Food

Holes in the Cheese: On Food and Cooking: The Science and Lore of the Kitchen, by Harold McGee (Scribner).

Forming a Skin: On Food and Cooking: The Science and Lore of the Kitchen, by Harold McGee.

The Telltale Fish: Joseph H. Hotchkiss, professor of food science at Cornell University.

Ginseng Pedigrees: Varro E. Tyler, emeritus professor of pharmacognosy at Purdue University.

The Physics of Noodles: Joseph H. Hotchkiss, professor of food science at Cornell University.

Chapter 5. Flora! Flora! Flora!

It's Easy Being Evergreen: Plant pathology department of Iowa State University extension service.

Unable to Let Go: A. Wayne Cahilly, manager of the arboretum and grounds at the New York York Botanical Garden in the Bronx.

Trees' Partners: Gerald Posner, an arborist with Arbor-Cultural Consultants in Brooklyn, New York.

Doomed Bamboo: American Bamboo Society.

Love That Salt: The Halophyte Biotechnology Center at the University of Delaware College of Marine Studies.

From Silk to Kernel: "How a Corn Plant Develops," a report by the Cooperative Extension Service of the Iowa State University of Science and Technology at Ames; Dale E. Farnham, assistant professor of agronomy at Iowa State.

The Dreaded Rhubarb: Rodney L. Dietert, professor at the Veterinary College of Cornell University and director of the Institute of Comparative and Environmental Toxicology.

Gaining a Toehold: Southern Wild Flowers and Trees, by Alice Lownsbery (Frederick Stokes Company).

Man-Eating Plants: Barry Meyers-Rice, conservation co-editor of *The Carnivorous Plant Newsletter.*

Chapter 6. All Bugs Great and Small

Bugs on the Windshield: Louis N. Sorkin, an entomologist at the American Museum of Natural History; James Jarratt, an entomologist at Mississippi State University; Michael May, an entomologist at Cook College of Rutgers University.

A Backyard Census: Kefyn M. Catley of the American Museum of Natural History.

The Swarms of the Annoying: The Natural History of the Mosquito, by Marston Bates (Peter Smith).

Aphid Husbandry: American Social Insects, by Charles D. and Mary H. Michener (Van Nostrand).

Lines of Communication: The Ants, by Bert Holldobler and Dr. Edward O. Wilson (Harvard University Press).

Butterflies over Broadway: The Monarch Program, Encinitas, California.

Fat for the Flight: David Marriott, founder of the Monarch Program.

High Flies: Elson J. Shields, professor of entomology at Cornell University.

Bites in the Sunlight: Associated Executives of Mosquito Control Work, New Jersey.

More of a Sting: International Toxin and Venom Data Base.

Long Live the Spider: American Spiders, by Willis J. Gertsch (Van

Nostrand); *The Black Widow Spider,* by Raymond W. Thorp and Eldon D. Woodson (Dover).

Source of the Flies: Rob DeSalle, the American Museum of Natural History.

Chapter 7. It's a Zoo Out There

How the Fittest Stay That Way: Pat Thomas, curator of mammals at the Bronx Zoo.

The Mighty Wolverine: Macmillan Illustrated Animal Encyclopedia; Encyclopedia of Mammals (Facts on File).

To Sleep, Perchance to Hibernate: Bears, by Wayne Lynch (Mountaineers Books).

Bear Pairs: The American Zoo and Aquarium Society.

Asleep on Their Feet: Katherine Houpt, a veterinarian who directs the Animal Behavior Clinic at Cornell University.

Calling a Snake a Snake: Darrel Frost of the Department of Herpetology at the American Museum of Natural History.

A Meal in One Gulp: William Holmstrom, collection manager in the Department of Herpetology at the Bronx Zoo.

Up a Creek with No Legs: Rattlesnakes, by Laurence M. Klauber.

Chameleon Colors: Macmillan Illustrated Animal Encyclopedia.

The Fertile Turtle: Grzimek's Animal Life Encyclopedia (Gale Group).

Frogs' Flypaper Tongues: Kraig Adler, professor of biology in the Department of Neurobiology and Behavior at Cornell University.

Flight of the Fish: Encyclopedia of Fishes: A Comprehensive Illustrated Guide by International Experts (Academic Press).

Telling the Dolphins from the Porpoises: Macmillan Illustrated Animal Encyclopedia.

Facts of Life for Whales: Gray Whales, by David G. Gordon and Alan Baldridge (Monterey Bay Aquarium Press).

Sex and the Single Fossil: John Krigbaum, an anthropologist at New York University.

The Species Census: Niles Eldridge, curator of the division of paleontology of the American Museum of Natural History; *The Diversity of Life,* by Edward O. Wilson (Belknap Press).

Chapter 8. Secrets of Cats and Dogs

A Cat's Mad Dash: Katherine Houpt, a veterinarian who directs the Animal Behavior Clinic at Cornell University.

Feline Antennae: James R. Richards, a veterinarian who heads the Cornell Feline Health Center at Cornell University.

Shocking Behavior: Alan J. Friedman, director of the New York Hall of Science in Queens.

A Kidney for the Cat: Clare R. Gregory, professor in the Department of Surgery and Radiological Sciences, University of California at Davis School of Veterinary Medicine.

Sensing a Storm: Lee Zasloff, associate director of the Center for Animals in Society at the University of California at Davis School of Veterinary Medicine; Dan Graf, a meteorologist at Pennsylvania State University.

Fido's Brain: Richard Joseph, a veterinarian and staff neurologist at the Animal Medical Center in Manhattan.

Garrulous Parrots: Timothy DeVoogd, associate professor of psychology at Cornell University.

Guinea Pig Identity Crisis: Michael Novacek, senior vice president and provost of the American Museum of Natural History and curator of vertebrate paleontology.

Chapter 9. Animal Aviators

Beyond Lungs: Manual of Ornithology: Avian Structure and Function, by Noble S. Proctor and Patrick J. Lynch (Yale University Press).

Sensing Something in the Air: Ornithology, by Frank B. Gill (W. H. Freeman); *The Cambridge Encyclopedia of Ornithology.*

Invasion of the Crows: Dr. Kevin J. McGowan, associate curator of birds and mammals at Cornell University.

Staying in the Pink: The Cambridge Encyclopedia of Ornithology; Ornithology, by Frank B. Gill.

Soaring into Old Age: The Audubon Society Encyclopedia of North American Birds (Knopf); *Manual of Ornithology: Avian Structure and Function.*

Turning on the Fly: Ron Rohrbaugh, an ornithologist at the Cornell Laboratory of Ornithology.

The Birds and the Bees and Birds: Ornithology, by Frank B. Gill.

Bird Bombs Away: Manual of Ornithology: Avian Structure and Function.

Where the Gulls Are: Ronald W. Rohrbaugh.

Hard-Heads: The Diversity of Life, by Edward O. Wilson.

Chapter 10. How Humans Get That Way

There Must Be a Reason: Margie Profet, an evolutionary biologist at the University of California at Berkeley.

The Advantages of Being Female: World Health Organization.

Boy Twins, Girl Twins: The Mothers of Twins Clubs; National Center for Health Statistics.

The Longest Digit: Randall L. Susman, professor of anatomy at the Medical School of the State University of New York at Stony Brook.

Limits to Growth: W. Ted Brown, a geneticist at the Institute for Basic Research on Staten Island.

Stroller Potatoes: Martin I. Lorin, pediatrician and author of *The Parents' Book of Physical Fitness for Children* (Atheneum).

Straightening the Spine: The Merck Manual of Diagnosis and Therapy; Stuart L. Weinstein of the Department of Orthopedic Surgery of the University of Iowa Hospitals and Clinics.

Chapter 11. Our Bodies, Our Aches and Pains

Manufacturing Blood: Robert L. Jones, president of the New York Blood Center.

Those Icy Fingers: Harry L. Bush, Jr., a vascular surgeon at New York Weill Cornell Medical Center.

Liver Spots: American Academy of Dermatology.

Those Aching Eyes: Daniel Wagner of the Sleep-Wake Disorder Center at New York Weill Cornell Medical Center.

Protecting the Ear: American Academy of Otolaryngology.

Not Just an Appendix: Carol Shoshkes Reiss, an immunologist at New York University.

Wrinkling Over Time: Larry Meyer, a dermatologist and geriatrician at the University of Utah Health Sciences Center.

Losing Color: National Vitiligo Foundation.

Damaging the DNA: Darrell Rigel, a professor of dermatology at New York University; American Cancer Society.

Putting the Pinch on a Nerve: The American Academy of Family Physicians Family Health & Medical Guide (Word Publishing).

Runaway Scars: The Merck Manual of Diagnosis and Therapy.

Too Much of a Good Thing: The Merck Manual of Diagnosis and Therapy.

Oldies but Goodies: United Network for Organ Sharing.

Final Changes: Anoka and Wright County Coroner's Office, Minnesota.

Chapter 12. Things You've Always Wanted to Know

An Icy Stare: Amilia Schrier, an ophthalmologist on the faculty of the Columbia Presbyterian Medical Center.

Sneezing After Hours: Gillian Shepherd, an allergist at New York Weill Cornell Medical Center.

Television Vision: American Academy of Ophthalmology.

Ticklish Subject: Sarah-Jayne Blakemore of the Wellcome Department of Cognitive Neurology at University College, London.

A Pain in the Side: Everett Murdock of California State University at Long Beach; *Running Injuries,* by Tim Noakes (Oxford University Press).

Catching Up on Sleep: Losing Sleep, by Lydia Dotto (Quill/William Morrow).

Nerves and Teeth: Daniel M. Meyer, head of the Division of Science of the American Dental Association.

No Stomach: Paul Basuk, a gastroenterologist at New York Weill Cornell Medical Center.

Chapter 13. Eat, Drink, and Be Healthy

All Pizza, All the Time: Marion Nestle, professor and chairwoman of the Department of Nutrition and Food Studies at New York University.

The Trouble with Mold: Donna L. Scott, a food scientist who is a senior extension associate at Cornell University.

Feeding the Nails: Hilary Baldwin, a dermatologist in Brooklyn, New York.

Priming the Heat Pump: Samir Zakhari, a pharmacologist who is director of the Division of Basic Research at the National Institute on Alcohol Abuse and Alcoholism.

Calcium Without Complaint: Food and Nutrition Information Center of the United States Department of Agriculture.

Tea for Therapy: The Honest Herbal, by Varro E. Tyler, emeritus professor of pharmacognosy at Purdue University (Haworth Press).

Colonizing the Gut: Kathryn Boor, assistant professor of food science at Cornell University.

Tempest in a Tuna Can: Rodney L. Dietert, an immunotoxicologist who is head of the Breast Cancer and Environmental Risk Factors Program at Cornell University.

Chapter 14. Battleground: The Body

Wash Your Troubles Away: American Medical Association.

Persistent Pathogens: John R. La Montagne, deputy director of the National Institute of Allergy and Infectious Diseases.

Long-Lasting Immunity: Regina Rabinovich of the National Institute of Allergy and Infectious Diseases.

Feline Fever: Article by Drs. Russell Regnery and Jordan Tappero in the journal *Emerging Infectious Diseases.*

Mad Cows and You: Peter Lurie, deputy director of the Public Citizen's Health Research Group and a member of the U.S. Food and Drug Administration's advisory committee on transmissible spongiform encephalopathy.

AIDS and Mosquitoes: Anthony Fauci, director of the National Institute of Allergy and Infectious Diseases.

Rodents and Rabies: Jean S. Smith of the Division of Viral and Rickettsial Diseases at the Centers for Disease Control and Prevention.

Missing the Bus: Patrick L. Kinney, assistant professor of public health at the Columbia University School of Public Health.

Chapter 15. Ways Modern and Ancient

Hot Stuff: Louis A. Bloomfield, professor of physics at the University of Virginia.

Fighting Fog: Ken Duffie, chief engineer for H. L. Bouton, Inc., Wareham, Massachusetts.

The Brain as 3-D Camera: Rachel Cooper, author of educational material for the Optometrists Network.

The Right Touch: Annette Burak, Microtouch Systems, Inc.

Exploding Elevators: Ralph Regan, safety director of the Federal Grain Inspection Service of the U.S. Department of Agriculture; Robert W. Schoeff, emeritus professor of grain science at Kansas State University.

Lightning Target: United Lightning Protection Association; National Lightning Safety Institute.

Burning Issue: Joe Nickell, a senior research fellow of the Committee for the Scientific Investigation of Claims of the Paranormal.

The Greeks Had a Name for It: Bert S. Hall, associate professor at the Institute for the History and Philosophy of Science and Technology at the University of Toronto; *A History of Greek Fire and Gunpowder,* by J. R. Partington (Johns Hopkins).

Letting It All Out: The Doctor's Dilemmas, Louis Lasagna (Harper).

The Awkward Amphora: Shelley Wachsmann, nautical and biblical archaeologist at the Institute of Nautical Archaeology at Texas A&M University.

Ancient Census: Tony Wilkinson, research associate at the Oriental Institute of the University of Chicago.

Digital Computing: From One to Zero: A Universal History of Nothingness, by Georges Ifrah (Viking); *The Nothing That Is: A Natural History of Zero,* by Robert Kaplan (Oxford).

The Mummy's Curse: Peter Dorman, former director of Egypt House, the Oriental Institute of the University of Chicago archeological installation in Luxor, Egypt.